A HEART AFIRE

HELEN BROOKE TAUSSIG'S BATTLE AGAINST HEART
DEFECTS, UNSAFE DRUGS, AND INJUSTICE IN MEDICINE

PATRICIA MEISOL

The MIT Press
Cambridge, Massachusetts
London, England

The MIT Press would like to thank the anonymous peer reviewers who provided comments on drafts of this book. The generous work of academic experts is essential for establishing the authority and quality of our publications. We acknowledge with gratitude the contributions of these otherwise uncredited readers.

This book was set in ITC Stone Serif Std and ITC Stone Sans Std by New Best-set Typesetters Ltd. Printed and bound in the United States of America.

Library of Congress Cataloging-in-Publication Data

Names: Meisol, Patricia, author.
Title: A heart afire : Helen Brooke Taussig's battle against heart defects, unsafe drugs, and injustice in medicine / Patricia Meisol.
Description: Cambridge, Massachusetts : The MIT Press, [2023] | Includes bibliographical references and index.
Identifiers: LCCN 2022061539 | ISBN 9780262048521 (hardcover)
Subjects: LCSH: Taussig, Helen B. (Helen Brooke), 1898-1986. | Pediatric cardiologists—Maryland—Biography. | Women physicians—Maryland—Biography. | Women cardiologists—Maryland—Biography. | Women pediatricians—Maryland—Biography. | Physicians—Maryland—Biography. | Cardiologists—Maryland—Biography. | Pediatricians—Maryland—Biography.
Classification: LCC RJ43.T38 M45 2023 | DDC 616.1/20092—dc23/eng/20230707
LC record available at https://lccn.loc.gov/2022061539

10 9 8 7 6 5 4 3 2 1

For the Catherines: my mother, who told me stories of how women make change, and my daughter, who is on her way to making change.

He is great who is what he is from nature, and who never reminds us of others.
—Ralph Waldo Emerson, "Uses of Great Men"

CONTENTS

PREFACE

This book began as a conversation between two mothers on the sidelines of an athletic field at the Bryn Mawr School in Baltimore. Elizabeth Nye Di Cataldo was writing a history of the 125-year-old girls' college preparatory school and gave me a book she had used in her research. It was so compelling that I cold-called the author, Kathleen Waters Sander, to thank her for writing it. *Mary Elizabeth Garrett: Society and Philanthropy in the Gilded Age* is the story of a woman and her circle of friends who used her great wealth to educate girls and women. They met on Friday evenings in Garrett's mansion on Baltimore's Mount Vernon Square, often described as one the most beautiful urban spaces in the United States, to plan how to advance women's freedoms.

I had always looked for stories of women who work in creative ways to free themselves and others, parallel stories to ones we know well of men who bring about institutional and cultural change. I had first heard such stories from my mother. As a journalist, I uncovered others. These were often single acts of courage by women. Sustainable change by women was harder to detect and to reveal because it often involves incremental acts and multiple actors, work that can be as tedious as it is powerful. Now it turned out there was one such story right under my nose. It took place on the street where I had once lived, though there is no sign of these women on the street today. Indeed, on the rainy day in March 1988 when I moved into a third-floor walk-up on what had become a neglected historic square, I saw only stone reminders of great men—railroad barons, philanthropists, an art collector, the author of the "Star-Spangled Banner," and, in the center of it all, a towering monument to George Washington.

Now, amid these monuments, I began to visualize women in long dresses rushing down the street, huddled in conversation, headed for Garrett's mansion. These were doctors breaking from work to report on their research and the obstacles they encountered. In the 1890s, when women were rapidly being excluded from science-based medical schools, Garrett used her money to win them a permanent place. In a deal with doctors at Johns Hopkins Hospital, Garrett offered to raise money to build the medical school they desperately wanted if this new school would set admission standards higher than any other in the United States and—significantly— admit women on the same terms as men. Once women arrived, Garrett and friends made sure they had their own room in the medical school and, after graduation, money for their research. She invited them to tea.

The medical school came after the girls' preparatory school, and about the same time Garrett and friends stabilized a new women's college in Pennsylvania, Bryn Mawr College, installing their intellectual leader, M. Carey Thomas, as president. Once a path to higher education for women was secured, Garrett financed and hosted what Sander in her biography identified as a pivotal moment in the battle for women's suffrage. This was the 1906 national suffrage convention in Baltimore, scenes of which again took place in Garrett's home on Mount Vernon Square. Among those she received were the aging Susan B. Anthony and Julia Ward Howe. They plotted to reinvigorate the movement's leadership and replenish their purses. Garrett raised the money for the next push.

It was stunning to think that the women *without* monuments on Mount Vernon Square helped free exponentially more people in our democracy than General Washington. By the time I learned of Garrett, women had been voting for ninety years. An alumna of Bryn Mawr College had just been named president of Harvard. What about the women who became doctors thanks to Garrett and friends? How did that part of the story continue?

This is the question I asked Sander. Over lunch, she told me about Helen Taussig.

*

The life of a children's doctor sounded dull, but tedium being where triumph unfolds, I decided to investigate by reading about Helen in the archives of the Johns Hopkins University School of Medicine. There, in her own hand, I learned about suffering by children that is unimaginable today and glimpsed the fervor she would use to map a new world: defects of the heart. As I sifted through these records, I noticed a portrait of Helen in the anteroom. I walked over and stared at it. Helen stared back. I stepped away, letting out a deep breath. Goose bumps told me I wanted to know this commanding lady. From an archivist, I learned that one of Helen's protégés still lived in Baltimore. With great anticipation, I knocked on the door of Helen's last living close associate, Charlotte Ferencz, an eighty-eight-year-old doctor who studied children's hearts with Helen in 1949.

For the next four years, we met regularly in Charlotte's apartment over tea. I would bring documents or stories I had unearthed that triggered Charlotte's memory and raised new avenues to pursue. She provided context to a historical time in medicine, the beginning of heart surgery, as well as details like what Helen wore. She challenged me to examine Helen's relationships with patients, which she, like others, considered over the top.

Charlotte had cared so little for the portrait that intrigued me that for nearly half a century she helped prevent it from hanging at Helen's alma mater. But she had always sought to understand Helen in all her complexities. In conversation, she described her mentor's oddities along with her charms. Gentle and scholarly, Charlotte maintained good relationships with the same people she criticized for their treatment of Helen. Her skepticism of Helen's contribution to preventing a thalidomide disaster and its aftermath led me deep into public records, books, and archived correspondence to document the precise role Helen played in drug regulation in America. It is reconstructed here for the first time.

Other meaningful events of Helen's life and work surprised Charlotte. Although Charlotte ran Helen's rheumatic fever clinic for four years, regularly sat with her at professional conferences, visited her in Cape Cod, and sought her guidance, she viewed Helen as an elder aunt rather than an intimate. Like so many others, she still addressed Helen by her formal, if shortened, title: "Dr. T." They had known each other nearly sixteen years the Thanksgiving morning when Helen begged Charlotte to address her like a sister. "Ever so much love, dear Sari," she ended her note, using the familiar version of Charlotte's Hungarian name, Sarlita. "Please call me HELEN!!"

Here, we will. Helen would have insisted.

Reisterstown, Maryland

1 THE WOMAN IN WHITE

It was early afternoon in May 1945, a few weeks before Charlotte Ferencz was to graduate from medical school at McGill University. She and the other women in her class (nine of ninety-four students) were the guests of honor at a tea party hosted by the women's medical society. Charlotte had taken her seat when the group's president raced into the room holding a magazine aloft in her hands. Had they heard the news? A surgeon in Baltimore had opened the chest of a child born with a defective heart—three such children's chests, in fact—and reordered some blood vessels to allow her to breathe normally. And all three had survived. Blue babies, they were called, because their skin reflected purplish blood short of oxygen. Now they were all pink.

A stunned silence followed. Then came the clinking of bone china. "We all put down our cups," Charlotte recalled. The new doctors jumped from their seats, leaning over the shoulder of the women's medical society president as she read the story of the three children in the *Journal of the American Medical Association*. The world of medicine they were entering had been upended. It was now possible to surgically change the way the heart works. Charlotte noticed that the first name on the article belonged to the surgeon, Alfred Blalock, of the Johns Hopkins University School of Medicine, and the second to a pediatrician there named Helen Taussig.

Thrilled though she was to see a woman taking part in such a revolutionary moment in medical history, it was not a name Charlotte recognized. A year later, Charlotte came across Helen's name again when she was asked to box up the papers of the late Maude Abbott, the Canadian doctor who first classified malformations of the heart from laboratory specimens she had amassed. In those boxes were several letters revealing that Helen consulted Abbott in the early 1930s before making her own breakthroughs—discovering how to identify these heart defects in the living from her patients' symptoms. Once she could diagnose them, Helen began thinking about how to fix them before they did their fatal damage. As Charlotte came to realize, the surgery that Blalock performed that saved these children's lives had actually been Helen's idea.

By the time Charlotte finished her residency four years later and won the chance to study children's hearts in Helen's Baltimore clinic, Helen's work with children abandoned as hopeless had made her world famous. A photograph of the smiling woman doctor dressed in floral prints, smart felt hat, and elegant gold pin, often accompanied by children, regularly appeared in the pages of *Le Monde*, the *New York Times*, the *Indian Express*, *Jerusalem Post*, the *Times* of London, *Life*, and Walter Scott's *Parade Magazine*. Patients arrived from around the world, some without appointments, hoping for a miracle.

Helen set the standard for a new medical specialty, pediatric cardiology. Inside her clinic and later as her close associate, Charlotte saw Helen engage in tenacious battles on behalf of her patients. She defied norms, not just for her gender but for what medicine itself was capable of. At her most relentless, such as the year Helen left her clinic to wage a national campaign against the drug industry, demanding regulation in the aftermath of its "miracle pill"—the sleeping drug thalidomide—Charlotte so tired of hearing about it that if Helen approached, she ducked for cover.

Helen had to be determined to do what she did. She was pigeonholed in a children's heart clinic because no male doctor wanted to care for babies destined to die. Then, immersed in the care of children, many of them harboring a highly contagious bacterium, came the whooping cough that would make Helen progressively deaf. But she was also generous and compassionate. This is how Charlotte and others saw Helen: with her dying babies, holding them, hugging them, talking to them long into the night, and then, with the saved ones, advising them over years, agonizing about ways to extend their lives. Her colleagues also witnessed the confidence her methods inspired in patients, parents, surgeons, and generations of children's doctors. They saw what she did for women doctors, ensuring their influence in medicine when technology, surgery, and drugs began to replace the doctor's touch.

It pains Charlotte that today this woman who made a difference in so many people's lives is forgotten outside academic medicine. There in her sunlit apartment, among her memories, Charlotte shares her own ideas of how Helen should be seen. But patients, those who experienced Helen, who felt her touch, are the best ones to ask, she says.

Their letters contain clues, revealing even, as one man wrote, that Helen motivated him to pursue a life that reflects hers: "I will never forget the tall woman dressed in white."[1]

Imagine a nine-year-old boy gasping for air, too miserable to eat, talk, or walk, subjected to test after test, expecting to die, wanting to die, blue from head to toe, feeling the hand of a lady doctor on his arm, alone with him as he falls asleep on a hospital gurney. Hours later when he awakes, she is still there, and he is breathing normally and his skin has turned pink and he feels fine—so fine, in fact, that he tries to get out of bed. The doctor is smiling, but her voice is firm and she holds him back. Not yet.

2   DIGGING FOR CLAMS

Helen Brooke Taussig would tell you there was nothing remarkable about her childhood. That was her characteristic modesty speaking. The youngest of four children and the one with the most severe scholastic challenges, she was born on May 24, 1898, into an elite society in Cambridge, Massachusetts. At that time, Boston was the center of American intellectual and civic life. Her father, a Harvard professor, was a preeminent theorist in the new field of economics, famous for his book upending the silver monetary standard. And in an era when university education was beyond the reach of most Americans, Helen's mother, Edith Thomas Guild, had studied biology at Radcliffe, the new branch of Harvard for women.

Edith Thomas Guild was Boston aristocracy—the families who led the American Revolution and now with their inherited wealth ruled the country—but she chose her friends for their virtues, not their social position.[1] Her fierce independence and willingness to question social norms fit with her childhood church too. King's Chapel in Boston, founded as the Protestant Church of England in the late 1700s, rejected the idea of the Trinity and transformed itself into the center of Unitarianism in America. The breakaway faith espoused reason, science, and public duty. Most genteel Bostonians, including the city's mayor and eight of Harvard's presidents, were Unitarian.

By Edith's teenage years, a revolt from within was in full swing. Americans questioned why they should rely on European philosophers like Immanuel Kant or even the Bible to interpret or reveal God. They hungered for something spiritual. The leader of the revolt was a Harvard man and a

minister's son, Ralph Waldo Emerson. In his writing and in talks throughout New England and later the country, he called for a renewal of spiritualism and a direct relationship with God. He based his spiritualism on
nature, which he considered divine. His view was that each person created
his life through a series of acts. This new idealism was transcendentalism.
Once again, Edith's relatives were front and center of what Oliver Wendell
Holmes called America's "second declaration of independence."[2]

Like Emerson, Helen's mother was captivated as much by the processes
of nature as by the truth and beauty they revealed. In another era, she
might have been a botanist who dispensed medicinal plants. Even after
the birth of her first child, William, Edith Taussig returned to Radcliffe to
study plant life. When the arrival of a second child, Mary, prevented her
from completing her university degree, Edith found other ways to indulge
her curiosity for nature, which included persuading her widowed mother
in 1896 to purchase a summer home in Cotuit. The tiny village of Cotuit
along the scraggly coast of Cape Cod would inspire and comfort Helen
throughout her life. The cedar-shingled Taussig Big House, as the summer
home in Cotuit came to be known, contained fourteen bedrooms, two of
them for maids, and faced Nantucket Sound.[3] Waves brought the saltwater
nearly up to the long porch.

*

Despite the pedigree of her birth, Helen's family in her early years was in
crisis. When Helen was born, her father was at the height of his career.
Even in summer, he worked several days a week, riding his bicycle nine
miles from the family house in Cotuit to the Boston-bound train in Barn-
stable.[4] Along with editing the *Quarterly Journal on Economics*, he served on
numerous state and federal committees on taxes, his specialty. Four children, teaching, a heavy writing load, and multiple public commitments
proved too much. In the summer of 1901 when Helen was three, forty-one-
year-old Frank Taussig suffered a nervous breakdown that would slow him
for four years.[5] Over two of those years, the family traveled to spas in Tyrol,
Switzerland, and the Italian Riviera, where Helen's father wrote nothing but
postcards to colleagues, including Cambridge neighbor William James, the
philosopher and psychologist, who offered tips from his own breakdown.

Helen's earliest memories, then, were of a swaying ship, her first sounds
of English mixed with foreign words. Helen's father would eventually make

a spectacular recovery from his breakdown, keeping his career by learning from a growing school of psychoanalysts and from his own father the importance of resting one's mind. As Frank improved and the family returned to their home on Scott Street, he took up his work again.

What a relief it must have been for Edith and the children to arrive back in Cotuit in April 1905. Six-year-old Helen and her siblings appeared "happy as birds to be on the Cape again after Taussig's long illness from nervous prostration," their neighbor, James H. Morse, wrote in his diary after the children ran over to visit him. The ground was covered in two inches of snow the April day he invited the family to an oyster roast. Helen came running ahead, barefoot.

Her tactile sense developed early. As a young child on the wet yellow sand, Helen gathered clams during low tide on the banks inside a channel created by a large spit of sand running along Popponesset and Oregon beaches. There, in front of the Morses' home, she would drop to her knees at the sight of tiny holes emitting squirts of water and dig furiously down a half-foot or more until she felt something soft—the siphon or neck of the clam—or sharp—its brittle oval white valves.

These creatures, with their malleable neck muscles containing tiny tubes to siphon water in and out, bringing oxygen and food, provided Helen with her first lesson on circulation. With practice, gently probing, digging, pushing, Helen became adept at knowing how much pressure to apply in search of the delicate clams whose shape she knew—so much so that decades later, when she could no longer hear the sound of a child's beating heart through her stethoscope, she could feel it with her hands.

The summer she turned eight, Helen's reading problem emerged. Her father, who as a teen sequestered himself in the attic to read Latin and Greek, encouraged his children to become voracious readers. Soon enough he noticed Helen's difficulty distinguishing words on a page. She had a form of dyslexia. Back in Boston for the school term, Helen quickly fell to the bottom of her class in reading and spelling. Other girls read aloud for pages until they stumbled over a word. Helen, faltering over the first sentence, got two chances before she was forced to return to her seat. Humiliating school days led to comforting evenings in her father's study. Frank Taussig did not give up. Nor did he raise his voice, punish her, or reveal his own frustration at her failure.[6] Instead, he offered praise and an alternative way of considering her dilemma. The reason she had a hard time spelling, he told Helen, was because spelling was not logical, and Helen was logical. With this acknowledgment of her intellectual promise, Helen worked harder to solve her word puzzles and, naturally, to please her father.

Around the same time, her mother introduced her to the sport of observing birds and plants. Edith encouraged Helen to learn the Latin names of flowers in the garden. Helen instinctively grasped the workings of living things, making up for her trouble with words. But botany lessons became sporadic as her mother's health waned. Beginning in 1907, Helen's mother fought sickness after sickness, occasionally recovering enough from flu-like symptoms to seem normal. Most people carried a latent form of the bacillus, *Mycobacterium tuberculosis*, the bacterium identified in 1882 by Robert Koch as the cause of tuberculosis.[7] But women weakened by childbirth could suffer mild symptoms on and off for years before the disease was diagnosed.[8] This was the case for Edith Taussig.

By the autumn of 1908, through the ritual late-August clambake and the October oyster roast Helen's mother and grandmother enjoyed with neighbors before they closed their Cotuit home, Helen could detect signs of trouble. Her sister, Mary, increasingly took on household chores and read to Helen. Her mother's cough worsened. Edith, accompanied by some family members, spent the winter of 1908–1909 in the resort town of Aiken, South Carolina, in the hope that the warmer climate would improve her health. Doctors said she had Hodgkin's disease, a cancer that attacks the body's immune system. To a ten-year-old child beginning to realize her own flaws and watching her mother's prolonged suffering, life could not have been hopeful. There is some evidence that her father shielded Helen and her siblings from the realities of their mother's illness and then from contracting it. Possibly beginning in the summer of 1909 in Cotuit, Helen's mother was isolated in a third-floor bedroom. Each evening, Frank sat outside Edith's door and kept her company.[9]

The family rituals continued. That July, Helen and her grandmother joined a crowd of families for tea in the Morse barn. On another day, she watched her father and other men gather for a "smoker" in the Morse barn to chat about one of their Cotuit neighbors—Abbott Lawrence Lowell—who was about to assume the presidency of Harvard. Helen was thrilled in September when the aunt for whom she had been named arrived from Boston. Helen Ellis was like a sister to Helen's mother, and her husband, the

Rev. Stopford Brooke, a Unitarian minister at First Church in Boston, was the son of the renowned Irish literary figure Stopford Augustus Brooke.[10]

By then, Helen's father recognized tuberculosis from the unrelenting progression of his wife's disease, which started in the lungs and caused the victim to cough up blood and mucus. The only treatment was good nutrition and fresh dry air. Abruptly he canceled the family's trip abroad and used his sabbatical from Harvard to care for Edith, finally taking her away to the preeminent sanatorium for tuberculosis, the private care cottages at Saranac Lake in New York's Adirondack Mountains.[11] Helen's brother, William, enrolled at Harvard, but she and her two sisters accompanied their father, staying in a rented house in a nearby village so they could visit their mother. It was a dismal time. The disease ate at Edith until she resembled a living skeleton. She died there in April 1910.

Helen was not yet twelve. Despite a remarkable memory, she rarely spoke of her mother to friends. Whether from fear or hope, long after modern medicine made the cottages at Saranac seem quaint, long after the discovery of a vaccine to prevent TB and antibiotics to cure it, Helen embraced the healing power of fresh air. From childhood on, she slept on a porch or near a window, some nights breathing air so cold that by morning, the water in her bedside glass had turned to ice.

The family returned to Cotuit to mourn. Helen's grandmother opened the house, but her sister Mary, age eighteen, ran it, showering attention on Helen. This was to be Helen's debut summer on the Cotuit skiff. Beginning in 1906 and still today, the social life of Cotuit children revolved around racing these flat-bottomed boats with oversized sails. With William back from Harvard, the four Taussig children sailed together. William won the season-ending Challenge Cup summer after summer except once when he capsized and, in a second year, lost a mast.[12] Catherine, too, bright, bold, and daring, was emerging as a champion sailor. Alas, the siblings' instruction as Helen raced a Cotuit skiff for the first time was not enough to convince her the thrill was worth the terror. An adolescent with so little control over the world would not want to be tossed and turned in the wind and waves.

Helen's father recovered from his wife's death by plunging into his work and seeking counsel from his sister, Jenny, of Louisville, Kentucky, with whom Helen would be close. To his children, Frank demonstrated self-discipline; every morning he rose early to write for four hours, not to be disturbed. He published in 1911 his *Principles of Economics*, which became the preferred textbook in the United States and England for four decades. But he was also devoted to his children and indulged their play, reserving summer afternoons to swim and play tennis with them and entertain them with his violin. His youngest daughter, more comfortable in a canoe than in a Cotuit skiff, and carrying the bacteria that killed her mother, needed him most. He obliged. Between Edith's death in 1910 and 1917, the year Helen entered Radcliffe College, Frank Taussig worked to make his daughter feel that it was the rest of the world, not his brilliant daughter, that was the problem. It was an opportune time for Helen's development.

Her father was an imposing man. Like his mother's father, a Lutheran schoolmaster, Frank Taussig was a teacher at heart and possessed an unorthodox mind. As a young man, Frank observed and rejected the lecture-style teaching in Göttingen and other top German universities; he believed in engaging students, and at Harvard he enjoyed a reputation as a great teacher. Each weekend the professor opened his home to graduate students, colleagues, and visiting scholars, setting a formal but welcoming table, and giving Helen an earful of provocative conversation on social and economic issues. One topic was so-called scientific management, a term coined by a Taussig relative by marriage, who would become one of America's greatest jurists, the Supreme Court justice Louis Brandeis. Boston was the center of management studies, notably by Frederick Taylor, who posited that the way to improve efficiency of industrial production was to break work into pieces and reward workers for their speed at completing the same job over and over. On top of that, monopolies pushed down wages. For families to survive, children had to work.

Unhappy workers in the coal, railroad, and meatpacking industries began to organize. Throughout Helen's childhood, the streets were full of marches and protests over working conditions and wages and, for women, temperance and the vote. In the aftermath of the 1911 fire at the bolted-shut Triangle Shirtwaist Factory in New York City that left 128 dead, debate at her father's table turned to worker safety and wages. Massachusetts in 1912 passed the first minimum wage for women and children, and civil leaders debated whether women should be paid the same as men for equal work. The professor didn't think so; he favored what the market would bear and argued so in his *Economics Quarterly* journal. A minimum wage would reduce employment of the lowest-skilled workers, he believed. At the same time, he shared the Progressives' worries about the power of monopolies.

Efficiency fueled America's emergence as a rival to England in a burgeoning global economy in the early 1900s and led to hotter debate still over the "drag" on society from so-called undesirables—people with physical or mental weaknesses. The solution that many Progressive economists advocated was eugenics, the practice of encouraging only the genetically superior to reproduce (about which Frank wrote, endorsing their ideas but discussing the pros and cons).

Helen also accompanied her father abroad to academic meetings, absorbing more such debates and, in the summer of 1916 while in high school, to the University of California, Berkeley. Berkeley, where Helen studied and played tennis while her father lectured, made a lasting impression. Women, including one of Helen's friends, were everywhere and conducted themselves equally to men. It had been coed for sixty years.

Her father oversaw her preparation for Radcliffe. Mary attended the Winsor School, the top Boston preparatory school. Helen, hampered by a reading deficit and weakened by a mild mediastinal tuberculosis, enrolled in the second-tier Cambridge School for Girls. Helen was so fragile she attended school only from 9:00 a.m. to 11:00 a.m. each day.[13] The assignments would have been arduous given her health, but they were more challenging still for a girl with reading difficulties—George Eliot's *Silas Marner*, Tennyson's *Idylls of the Kings*, *Macbeth*, Milton, and American poets. With her father's help, Helen found alternative ways to read, often by looking at sentences from their endings to beginnings. As she experimented with ways to unlock a sentence or a paragraph, she discovered patterns that she committed to memory.

By sophomore year, she was allowed to stay at school until 12:30 p.m. Other sickly children required to rest during the day entertained themselves by reading. Helen developed shortcuts to reading. First, she discerned what she needed to know to master ancient Rome, Latin, German, and algebra. Then she asked questions about each book to find out if it would help her learn. This proved economical: she did not waste effort on things of little value. The habit of investigating and selecting her reading carefully would become a major asset. So would her propensity to review her answers. Her father insisted she check her work three times. This way, he explained, she would be sure of her facts. With this system of identifying patterns, questioning, and review, Helen concealed a lifelong handicap. She could not write six numbers without reversing one, but the endless lists she would record for her research were as accurate as those of any other scientist. Letters to friends would reveal no evidence of the struggle to write them. This required her to work harder than others and to persevere. Her father abetted this, challenging her, point by point, match by match, on the tennis court behind the Taussig Big House.

Bed rest, nutrition, and exercise checked Helen's latent disease and strengthened her willpower. By the time she graduated from high school in spring 1917, she possessed a dominating forehand and a killer tennis serve. She was also cheerful, a leader among her schoolmates, in possession of an excellent mind and, according to her college recommendation, in perfect health.[14] One of her strengths was physics. In her junior year, she performed thirty-seven experiments. Armed with a strong will, she was ready to fend for herself.

The timing was ripe since her father was in demand, notably by Woodrow Wilson, then the president of the United States. Wilson had asked Congress to declare war against Germany, which had been menacing Britain and its allies for three years with nerve gas and a fleet of submarines. The question of how to finance the war loomed, and Wilson sought the professor's advice during a prolonged spring trip to Washington. Helen's father was critical of taxes on ballooning global trade, but he was practical, not doctrinaire. That made him the ideal candidate to chair Wilson's new US Tariff Commission, which the president hoped would find a politically palatable way to finance the country's entry into World War I. Helen's father agreed to move to Washington. For a few days, the family gathered in Cotuit, where, Morse recorded in his diary, the "merry voices of young ladies" once again filled the pinewood path between the Taussig Big House and the Morse barn.[15]

In Washington, Helen's sister Mary had not yet finished unpacking and preparing her father's new quarters when a letter arrived from Helen in which she complained that she was bored. To her father's surprise, Helen asked his permission to transfer to the University of California, Berkeley.

California was a four-day journey by train, but the ambitious and adventurous Helen had discovered the Harvard faculty lent to Radcliffe were either away in Washington to help in the war effort or uninteresting. Helen was also weary of the expectations people had for the daughter of a

Harvard professor. Cambridge was a small town that thrived on tradition. Berkeley, with 8,000 students, was full of opportunities. Plus, several friends had enrolled.

Frank Taussig tried to reassure his daughter. "Dearest Girlie," he wrote from the Hotel Grafton in Washington, using the German term of endearment. "It is natural that Radcliffe should start slowly."[16] Vacancies were the norm on campuses all over the country, he told Helen, and while her professor would take getting used to, he would teach Helen to think. Frank would know; he had helped convince Harvard faculty to teach part time at Radcliffe.

Loneliness trumped Helen's loathing of writing; she took up a pen every day to describe her longings to her sister Catherine at Bryn Mawr College. To her father, Helen complained at length about difficulties in English class. From the flow of her pen, he saw its benefits.

Visiting Washington the following summer, Helen found her father preoccupied not only by world events but also by the warm and much-admired founding director of Boston's public kindergartens, Laura Fisher, who was tied to the Taussig clan by marriage. That August, "Cousin Laura" accompanied the Taussigs to Cotuit. The days of summer frolicking were few, however. A German U-boat sank three coal barges in Nantucket Sound, and the next day, an American hydroplane skimmed around the Taussigs' house like a mighty bird. On August 31, the professor, then fifty-eight years old, stopped at Morse's hammock to announce his wedding that day to Miss Fisher. His daughters also stopped by. "Fine looking ladies they now are, Mary, Catherine, and Helen, all grown up tall, Helen particularly tall, all of them beyond the oyster roasting days," Morse wrote.[17]

Academic progress was the condition her father imposed on Helen when she persisted in her quest to transfer to Berkeley. Her success during a second year at Radcliffe in the face of loneliness, war, and tragedy boosted her confidence. There is no record of letters between father and daughter after a deadly strain of flu carried in by sailors in the Boston harbor began ravaging the city. By October 1918, two hundred people a day died from influenza (Spanish flu); on Boston streets, body after body was loaded into a Red Cross wagon. Some weekends, Helen and Catherine escaped to Cotuit. Her father, representing the United States at the Paris Peace Conference, surely worried about his daughters. All we know is that when again Helen expressed her desire to move across the country for her junior year, he agreed.

Helen's siblings chose spouses from among Frank's protégés or within his academic orbit. She journeyed by hot, dusty train to California. Helen's father found her a room in the home of a Berkeley English professor, Chauncey Wells, who was married to "Aunt Mary"—Mary R. Prescott, a relative of Helen's mother. It was the perfect setup, providing Helen with a built-in editor for her term papers. By Helen's account, she was treated as their daughter. She studied by the parlor fire and tended the flower garden. At every opportunity in her last two years of college, Helen explored and amused herself. Photos show her on stage in Greek regalia, skating, and rowing crew. Over Christmas vacation, she hiked and camped with classmates through Yosemite National Park, where a youthful Ansel Adams was photographing vistas that would introduce Yosemite to the world. She went to the Rose Bowl.

Driving back to California for her senior year, Helen and a friend stopped in Denver, explored the Rio Grande, and photographed San Francisco in its smog. It was a crowning year. Helen captained the women's tennis team to championships. She made headlines in the *Daily Californian* for donating $50 ($1,300 today) earmarked for extras, not operating expenses, to support the record number of women students, fifteen hundred, who joined sports teams. The Phi Beta Kappa honor society on her diploma in May 1921 marked her as among the brightest the country had to offer. At the age of twenty-three, Helen was serious and mature. Like her father, she was slender and tall, with a swimmer's arms. She shared his rounded cheeks, a full lip, and love for music. She was equal to him in another way too. The last state had ratified the Nineteenth Amendment to the US Constitution in 1920. Helen could now vote.

When she returned home after graduation, Helen began to seriously consider enrolling in medical school. A classmate had suggested it after witnessing her success in physics. Several friends were already on their way to becoming doctors. Frank Taussig would soon be home; he had asked President Wilson to relieve him of his federal duties because Laura was ill.

Dancing under the stars one evening with Catherine and two friends in Cotuit, Helen celebrated the birth of Mary's first child and her own yearning to explore the world, a world whose joys could easily devolve into suffering without vigilance. Her struggle with words had hardened her determination and given her the ability to anticipate events based on the facts she assembled through the wide-angled lens of her mind. She had demonstrated this a few years earlier in Cotuit. Lingering at her open window at midnight, gazing down at the pine needle path her friends had just traveled back to their homes, she had spotted a flickering light. Shouting to Catherine to find a mop, she ran downstairs to grab a fire extinguisher. The brushwood fire had spread twenty feet on either side of the path toward the Morse barn before Helen, aided by a new fire department, doused it.[18]

With such fearlessness, Helen asked her father for help obtaining a seat for her in Harvard's medical school when he returned to Boston. She had wanted him to use his influence as a prominent activist in university governance, but he deemed it impossible: Harvard had rejected every qualified woman applicant since 1847. Helen listed her reasons for medical school, including her success in physics. She was an experienced problem solver too, and in the news that summer, there was great excitement over an advance against the disease that killed Helen's mother. In France, doctors had just tried a new vaccine against tuberculosis.[19] There was so much to be done, and Helen was ready to jump in.

Helen's father recognized his daughter's talent, and he had supported nineteenth-century victories brought by suffragists that opened scientific medical schools to women. But Frank was a practical man and understood the obstacles Helen faced. Admission to medical school in the 1920s did not necessarily lead to a successful career in medicine. True, during Helen's college years, thirteen more coed medical schools opened, and women accounted for 5.8 percent of medical students.[20] The decision by elite medical schools—the University of Pennsylvania, Yale University, and Columbia among them—to admit a few women was business—a bid during World War I to bolster tuition hurt by enlistment in the armed services. Columbia made women pay for upgrades to accommodate them. At Yale, a professor financed an extra bathroom to ensure admission of women including his daughter. Once their financial position improved, these coed schools slashed spots for women. Medical schools didn't want women. It was most evident at Frank's beloved Harvard.

Lowell, the current president and the Taussigs' Cotuit neighbor, adamantly opposed admitting women, though he was willing to wring tuition from them. Women could audit courses—or some of them—in the medical school. Anatomy was off-limits, since discussing the body in mixed company was thought to be indecent. To stymie midwives competing with a growing number of male doctors delivering babies in hospitals, Harvard barred women from the two classes where they had a historic edge: obstetrics and gynecology.

*

In the centuries-old history of women healers, there was no era more challenging than the one Helen proposed to enter; the barriers to entry were steep: tough new medical school entrance requirements for everyone, far fewer spots for women, and, once in, few or no chances for women to obtain internships or advanced study required to work at the top tier in medicine.

The rise and fall of women physicians can be charted against cultural and social forces put in place by men since ancient Greece and Egypt. In colonial America, women were the predominant health care providers out of necessity. By the middle of the nineteenth century, when Helen's grandfather immigrated to St. Louis and began medical training, American women could earn a good living delivering care in rural communities and big cities alike. They delivered babies and studied plants for natural cures. Women's medical colleges opened, teaching midwifery and homeopathy. Botany for its medicinal forms was a revered science. Boston, home to the first medical school for women, the New England Female Medical College, was a hub for women doctors. By the end of the nineteenth century, 18 percent of the city's doctors were women compared with 5.6 percent in other cities.[21] By 1900, a few years after Helen was born, there were 7,387 female physicians in America, compared with fewer than 900 in England and 95 in France.[22] Between 1890 and 1910, the number of women physicians nearly doubled, but the pipeline was closing. Boston would not see such a high percentage of women doctors again until the end of Helen's life, in the 1980s.[23]

A handful of developments in the nineteenth century ensured the large-scale exclusion of women from practicing medicine in the twentieth century. First was the realization that medicine had become professional; men too could earn a living from it. Instruments such as the stethoscope made medicine more precise. In France, treatments began to be evaluated systematically, to see which ones worked. Hospitals, for centuries charitable places where religious orders and women reformers cared for the sick, poor, and mentally ill, began their transformation to places where people paid for treatment—to have surgery, to have babies.[24] American men sought training in Paris in the 1830s and later, in the 1860s, in Berlin and Vienna, and returned to set up medical schools to teach the new methods. Bacteriology and the study of diseases took off in the 1890s. Men who modernized the profession in the Victorian era intentionally excluded women. They believed that women's intellectual, emotional, and physical makeup made them incapable of doctoring in the new scientific methods, and that women would disrupt male students and lose interest in medicine once they married.

The British-born Elizabeth Blackwell proved them wrong, of course. The first woman to demand admission to a science-oriented medical school in the United States, graduating at the top of her class from the Geneva Medical School (later the State University of New York) in 1849, she obtained advanced training in Paris and returned to develop a network to help women in the profession. She opened her own hospital, the New York Infirmary for Women and Children, to give women the internships denied them by medical schools that were critical to practice and jobs. Beginning in the 1860s, Blackwell and others campaigned for science-based medical schools to admit women. Women won admission in 1870 to the University of Michigan and soon after to the University of California, San Francisco. Mary Putnam Jacobi, daughter of the Boston publishing family whose members were among Helen's summer friends in Cotuit, gained admission to the most respected medical school in the world, l'École de médecine in Paris. Jacobi, who flashed her checkbook to win entrance to medical research facilities normally closed to women, would popularize the idea of women buying their way into men's medical schools.[25]

In the United States, committees of suffragettes sprang up to try to pry open medical schools. Besides money, they used connections, newspapers, and lobbyists. As secretary to the venerable Harvard president Charles W. Eliot, a job he held while earning his doctorate, Helen's father witnessed the public debate over the fitness of women to be doctors in the 1880s that led Eliot to accept and then return a $10,000 donation from a group of suffragists conditioned on admitting women to the medical school.

Their biggest victory was at the Johns Hopkins University in Baltimore. There, in 1890, a group led by Mary Elizabeth Garrett, the daughter of a railroad magnate, and four other women, most of them daughters of Hopkins doctors or trustees, succeeded in opening permanent spots in medicine for women. The circumstances were different from Harvard, which already had a medical school. In Baltimore, a handful of European-trained doctors lured by the university had a new hospital, thanks to the will of Johns Hopkins. But when the value of his stock dropped, they lacked funds for the medical school he envisioned.

Garrett offered to raise an endowment, ultimately donating more than $350,000 of the $500,000 herself, on condition that women be admitted on the same basis as men.[26] (Among the next biggest donors were Boston women disappointed with Harvard.) Influenced by Martha Carey Thomas, who had been denied the chance to earn her doctorate at Hopkins, Garrett also negotiated admission standards to match those at top research universities in Germany. They exceeded any then required by an American medical school: an undergraduate degree, a reading knowledge of French and German, and a basis in the sciences. In the United States, a college degree was rare. Requiring one was risky for the doctors: it would limit the pool of male students and therefore tuition. To Garrett and friends, it ensured opportunities for graduates of women's colleges. With no other source of capital, the doctors accepted the deal.

When the Baltimore medical school opened in 1893 with three women and fifteen men, women accounted for more than 10 percent of students at eighteen medical schools teaching scientific methods.[27] On some campuses, more than 30 percent of students were women. With most women—66.4 percent—choosing coed medical schools, women's medical colleges, including those teaching the new methods, closed or merged at a fast clip.[28]

Then came the backlash. Women performed so well in science-based schools that the men complained. Fearful of losing them to the competition, medical schools cut back on admitting women students once their finances stabilized. The University of Michigan, where many college-educated Boston women enrolled to earn medical degrees, cut back from almost 19 percent women in 1893 to just over 5 percent in 1907.[29] Cornell, which had opened in 1899, accepting seventy women after a merger of Blackwell's women's medical college, had only ten women by 1903.[30] Against these attempts to marginalize them, women doctors formed their own medical association and started a journal.

The final step in the exclusion of women from medicine was the unintended consequence of one of the most influential reports ever made on higher education in America, the Flexner Report. Its goal was to promote science in medical schools and establish common admission standards. At the time, anyone could obtain a doctor's license—in some states, by giving a seminar. There were raucous disputes over standards, too many doctors for most to earn a living, and not much respect for them.[31]

In his 1910 investigation, funded by the new Carnegie Foundation to improve education, Abraham Flexner, a prominent educator, revealed that many schools had few teachers, little equipment, and poor facilities. Some lacked admission criteria or minimum academic standards. He recommended that medical schools model themselves after the Johns Hopkins University School of Medicine and seek a stamp of approval from a new regulatory body. Flexner's report transformed the teaching of doctors and American medicine. Medical schools became graduate research institutions. To be admitted, one needed a college degree and a background in physics, chemistry, and biology. Every student, even those who intended to practice in the community rather than conduct research, had to study laboratory sciences and learn to treat patients in an affiliated hospital.

Hiring full-time faculty, building laboratories, and forging alliances with hospitals required raising endowments and charging tuition that working-class students could not afford. Two-thirds of medical colleges closed, including the few remaining scientific-oriented women's colleges and those that trained Black doctors. Flexner assumed, mistakenly, that modernized schools would follow Hopkins's practice of admitting women. Instead, coed medical schools ultimately limited women to about 5 percent of the class, from fear that too many women would lead the small pool of college-educated male students to enroll elsewhere. Even Hopkins reduced women students, but it had a written agreement to admit them and a group of women determined to enforce it.

*

This was the backdrop for the conversation Helen had with her father on their porch in Cotuit. Doctoring in the community had nearly ruined her grandfather (William Taussig gave up his medical career due to a prolonged illness, later to become a much-admired St. Louis businessman).[32] Academic medicine was even harder. Even with medical schools setting aside seats for women, how many jobs would be open to them in a field commandeered by men? Frank Taussig knew how Harvard Medical School treated its only female professor. He rarely saw Alice Hamilton; nobody did, even though they hired her because she invented the field of industrial health, identifying poisons in the workplace and getting laws passed to protect workers.

Harvard did everything possible to make her existence clandestine. In her contract, Harvard forced Hamilton to agree never to eat in the faculty club, never to walk in an academic procession at commencement, and never, ever to expect football tickets.

The way things were unfolding for women, Helen might find far more opportunities in the booming field of public health, her father suggested. Indeed, public health research was transforming Boston; chlorinated water had eliminated typhoid fever and, despite recent heat-related deaths, reduced infant mortality. The city health department was systematically testing milk, and even ice cream sold on the street, for bacteria. That July of Helen's homecoming, city inspectors threw out hundreds of pounds of tainted cheese. Children were about to be tested for diphtheria, which could be treated if it was caught early enough. To prevent communicable diseases, nurses roamed into neighborhoods to teach sanitary practices. So many babies died of dehydration in summer 1921 that the city's Baby Hygiene office stayed open late to teach mothers proper techniques. Nurses and social workers were at the forefront of these unparalleled efforts to prevent disease and death. This work was so important that Harvard would soon open a new school for public health, modeled after one opened in 1916 at Hopkins. Alice Hamilton was moving from the medical school to become its star professor. With so many women employed in the field, surely this new school at Harvard would admit women, Frank told his daughter.

Both were shocked by what Helen discovered in a visit late that summer with the acting dean of this new school. Milton J. Rosenau, whose research led to pasteurizing milk, told Helen that applicants needed to study medicine for two years to be admitted. But even after studying medicine, women could only audit classes.

"Who is going to be such a fool as to spend two years studying medicine and two years more in public health and not get a degree?" she asked him.

His reply: "No one, I hope."

"I will not be the first to disappoint you," she said.[33]

3   HEARTS IN THE BATHTUB

That fall, 1921, Helen won special permission to audit a bacteriology class at Harvard and renewed herself on weekends in Cotuit with Catherine. After Christmas, her father took leave from teaching, and the sisters traveled with him and their stepmother, Laura, throughout Europe, staying well into summer 1922. But if their grand tour was intended as a distraction, a way for Helen to forget about medical school, it failed.

In lectures in histology at Harvard the following fall, Helen had to sit in a far corner of the amphitheater, away from male students, because she was a woman. In the laboratory part of the class, she cut and pasted tissues of plants onto slides and examined cells under a microscope, but in a separate room. "I was scarcely allowed to speak to men in the class for fear I would contaminate them," she remembered.[1] Still, the daughter of an esteemed Harvard scholar got special attention. Helen's professor came by her station daily to review her work. At the end of the semester, he told her not to waste time auditing classes and instead enroll in an anatomy class at Boston University's homeopathic medical school, which had absorbed the New England Female Medical College, where she could obtain credit that counted toward a medical degree.[2]

There, in 1923, Helen won an important ally. Alexander S. Begg, dean of the medical school and a highly regarded professor of anatomy, had been recruited from Harvard to preside over the transition of Boston University into a scientific medical school. Seeing Helen was advanced beyond his anatomy class, Begg gave her an ox heart to dissect and suggested she learn the muscles of the heart. He also gave her an article written by Franklin P.

Mall, a noted professor of anatomy at Johns Hopkins medical school. She did not understand it. With her hands, night after night in the laboratory in Boston, she tore apart the animal heart, trying to isolate its parts and intuit its workings on her own. "Mauled" was the word she used to describe what she did in her quest to understand how it worked.[3] She ruined it and, she feared, her chance to be a doctor. But instead of showing her to the door, the kindly Begg urged her to try again. He also told her she had progressed beyond most other students. Her spirits lifted. This was the moment when Helen decided to devote herself to the heart. Every evening thereafter, she could be found in the lab.

With the second heart came a new approach. This time, Helen dissected the mass of muscle that seemed so difficult to pry apart after first considering how it worked. To her surprise, the strips of muscle she had torn from the heart's ventricle and deposited in jars of solution suddenly began to wiggle. The heart of a mammal had been made to beat, and strips of ventricle attached to heart vessels. But no one had ever observed strips of muscle from the heart's lower chamber (ventricle) spontaneously contract. The experiment led Helen to publish her first academic paper, original observations on muscle contractions of the heart, in 1924.[4]

A published paper by a first-year student at any medical school was unusual. Begg suggested Helen was a better match for a medical school that emphasized scientific research. He recommended Johns Hopkins, where Mall had taught anatomy, and it had the most experience with women students. A recommendation from a Harvard medical school professor would carry more weight than one from him. Did she know someone?

Helen made an appointment to see her father's friend, Walter Cannon, the eminent Harvard professor of physiology, a specialist in shock and a member of the National Academy of Sciences. In his office, Helen explained her research and "very timidly" asked whether he might write her a reference.[5] To her surprise, Cannon knew all about her first experiment. We don't know if Frank Taussig had intervened with Cannon or sought his assessment before encouraging Helen, but he knew the details of her experiment well: he had edited her paper.

Cannon, John L. Bremer, the histology professor at Harvard, and Begg were among a cast of male academics and colleagues who helped Helen succeed in institutions whose structures and environment reflected a belief that women were inferior to men and did not warrant the same respect or job opportunities. If Harvard allowed women, he would recommend Helen there, Cannon wrote, "but since Hopkins is more liberal, I hope you will admit her." For Frank Taussig, Helen's departure from the family home would be wrenching. But he now unequivocally supported her decision to enroll in medical school. His support was crucial on another front, too. He paid the bill.

*

Baltimore, a shipping and railroad hub with a population of 733,826, was similar to Boston in size and industry in 1924 when Helen began medical school. It had a fledgling symphony, a baseball team, and the country's first public medical school, the state-chartered University of Maryland (1807). Like Boston, it had a thriving class of immigrants, more Germans than Irish, significant poverty among "Negroes," and a moneyed class. The affluent classes produced debutantes in the rural Greenspring Valley and a bohemian intellectual culture downtown. It included the Cone sisters, Etta and Claribel (also a Hopkins medical school graduate), who were assembling the collection of modern French paintings that today resides in the Baltimore Museum of Art. It also had Goucher College, an early women's college that prepared women for work in the sciences, and the legacy of Mary Elizabeth Garrett. For more than forty years, Baltimore had been benefiting from her gifts to improve the education of girls and women, beginning with the opening in 1885 of the Bryn Mawr School, a girls' college preparatory school. There was a path now for young girls and women who aspired to medical school.

Many women had distinguished themselves by the time Helen arrived at Hopkins. They included Margaret Handy, the daughter of a congressman, a 1916 graduate, and the first pediatrician in Delaware, who would gain fame for setting up a bank for mother's milk. Another was Martha May Eliot, a 1918 graduate who became the first director of the US Children's Bureau. Most remarkable of all was the student who, upon her graduation in 1900, had been invited to join the Hopkins medical school faculty despite its policy against women professors. She was Florence Sabin, professor of anatomy, a woman so brilliant that the spring after Helen arrived in Baltimore, the men who ran the National Academy of Sciences made an exception to admit her. She launched her career with a path-breaking book on the nervous system of the brain. Helen would not study with her, since she had taken anatomy in Boston, but she made a point of getting to know Sabin.

There were twelve women in Helen's class, making up about 10 percent of the students. She lived in a boardinghouse at 800 North Wolfe Street, a few blocks south of the hospital, with six other female students. Purchased in 1920 by the women's medical fund at Hopkins, the "Club," as the women called their house (the men called it the "Hen House") had eight or nine bedrooms, communal bathrooms, a kitchen and a hired cook, and a parlor where the women retreated after dinner. Those who lived there (or in a later iteration) describe it as a place that allowed them to form deep friendships and in nightly banter fend off sexist attitudes of male professors by laughing over their insults. Serious though they were, studying at all hours, on most weeknights after dinner, the women in Helen's era moved into the parlor and treated themselves to a quick game of Ping-Pong or a hand of bridge.

Helen was conspicuously absent. A housemate would marvel that she had not seen Helen at the bridge table once in the three years she lived in the house. Helen slipped away after dinner and headed upstairs, to work.[6] Her daytime schedule was crammed. Students attended lectures but were left to their own devices to master the material for final exams. Between lectures and the library, Helen hurried to the physiology lab. On her own, with animal hearts, she tried to repeat the experiments that she had done in Boston for her published article. Returning to the laboratory one evening to try again, she was surprised to find the doors locked.

That left only her boardinghouse if she wanted to repeat her experiments. She would have had to procure her own cow hearts too. Someone less exacting about these hearts might have walked south to the butcher at the popular Broadway market. But Helen's evening's dissection was probably delivered, preserved in ice, to her specifications. Such deliveries were common now that Baltimore's German butchers had been forced by city health officials to leave the fly-invested neighborhood known as Butchers' Hill and move to the country. In winter, a fresh cow heart immersed in cold water might last a week. Helen stored hers in the bathtub, under a dripping faucet.

Her housemates accommodated Helen, whose soft voice and New England manners were hard to resist, although her heart storage sometimes collided with their personal hygiene needs. Guests particularly were inconvenienced. None was more annoyed than alumna Handy, the Delaware pediatrician, who returned to the women's boardinghouse in 1925 for a week's stay while she underwent treatment for a broken leg. She was shocked to find herself waiting one night to use the second bathroom because Helen's hearts occupied the first one. Handy had opened a practice in Wilmington, going house-to-house to care for neighbors dying of the Spanish flu. She regularly made the long drive to Baltimore to sit in on seminars on the new specialty of pediatrics. But Handy could not imagine the fervor that compelled someone to dissect cow hearts at home.

Helen Taussig is "kooky," Handy told the other women. "She'll never amount to anything."[7] Handy would tell this story about herself and laugh after Helen's work with hearts made her world famous. But in the early years, Handy wasn't alone in thinking Helen was wasting her time, going off in directions that held little appeal. What could she learn, and why? Whatever would she do with it?

Helen only knew that to be successful with her studies of the heart, she had to understand how it worked better than anyone else. She was positioning herself for a residency in the most prestigious department, internal medicine. There was only one such residency for a woman in Helen's graduating class, which would be doled out based on her final standing.

The tub was not a tenable venue for her nightly research. A line for the bathroom on her account would have embarrassed Helen. At her height, just under six feet tall, she would have found it painful to bend over a tub for hours.[8] In a search for an alternative space, Helen appealed to Sabin, the lone female professor at Hopkins. Sabin found a place for Helen to work, precisely where is unclear from their letters, but there, again and again, Helen repeated the experiments that she had described in her published paper.[9] She had to be right.

These were only cow hearts. Possibly also with Sabin's help, Helen soon apprenticed herself to Edward P. Carter, a warm man who welcomed Helen into his cardiographic laboratory, where he studied abnormal rhythms of the heart. For the rest of medical school, she worked in his lab, examining the defects in autopsied hearts.

*

Helen experienced none of the overt hostility from male students she had felt at Harvard. At Hopkins as at the University of California, men and

Helen Brooke Taussig in 1927, the year she graduated medical school. Courtesy of the Alan Mason Chesney Medical Archives of the Johns Hopkins Medical Institutions.

women studied together. Her coed study group included three other students whose last names began with T, including Vivian Tappan, who was Helen's chief academic rival. They were known as "the three Ts." At the height of the Jazz Age, when women cut their hair, danced, smoked, and drank champagne, Helen in her long skirts, V-neck blouse, and string of pearls stood out for her "refinement and reserve," one of Helen's study partners, Harry Veach, would recall. With her regal height and precise manner of speaking, raising her eyebrows to emphasize a point, Helen could seem intimidating. But she also stood out for what Veach called her "kind, humanitarian eyes."[10]

But while the women played cards or studied into the evening, their male counterparts—and sometimes faculty—drank and played poker in private medical student fraternities. Prohibition (1920–1933) only slightly curtailed some of their late-night partying: they smuggled in alcohol gathered at a speakeasy down the street.[11] These clubs and their antiwomen atmosphere would continue through most of the twentieth century.

The ban on dating and marriage by medical students and interns had only recently been lifted after a professor fell in love with his student. In practice, though, little had changed: women studying to be doctors faced difficult choices about their personal lives. Some married, often to other doctors who encouraged their intellectual pursuits and continued to work. Women physicians were far more likely to marry than women lawyers, professors, nurses, or social workers, in fact.[12] But most women doctors in Helen's era, two-thirds of them, did not marry.[13] There wasn't an easy way to do both. Helen made a calculated choice: she knew she could be a good doctor but was not as sure she would be good at marriage.[14] The focus she required and the standards she demanded of herself for her profession left little time to devote herself solely to another person.

The social activity we know Helen allowed herself was a speaker series at the Club. She organized it, showing she was not only aware of women's issues but also that she was seeking to be among the next generation of leaders. One of the featured guests was Sabin. For a talk in 1925 or early 1926, she invited birth control advocate Margaret Sanger. Sanger's New York clinic was finally open, five years after she had been arrested for attempting to open it, and more clinics would follow, including, in 1927, one four blocks from Helen's boardinghouse. When Sanger's organization later tried to recruit Helen to provide birth control information at a booth at the Baltimore convention center, Helen demurred, explaining she was recovering from the flu after a stint through the maternity ward. Instead, she posted the request for volunteers on a medical school bulletin board.[15]

The truth was that she judiciously selected her activities, knowing that every second away from her studies would count against her. Along with voluminous reading and lab work to prepare for final exams, Helen needed to improve her German. The summer after her first year in Baltimore, Helen's father arranged for her to study German in the home of an economist friend, Gerhart von Schulze-Gaevernitz, at the University of Freiburg. Under no circumstances was Helen to hear a word of English, Frank wrote his colleague. While it would be a "deprivation" to both since Helen's German was rudimentary, Frank insisted Helen be allowed to "stumble along as best she can."[16] When Helen stepped off a steamer in Breisgau in early June to find her own way to the professor's home, she also carried with her letters of introduction to the university's top doctors. Helen intended to call on these experts to learn something new and share her own ideas.

Not until eight weeks later, when her father arrived, did Helen relax. The pair hiked leisurely that August in the Bavarian Alps. From there, they returned briefly to Cotuit before Helen's second year of medical school.

These interludes with nature were her only rest. When Helen returned to Baltimore, she was keenly aware that everything she did for the next two years was in preparation for final exams that would define her future. Then, burrowed in her room one week before exams, an unexpected phone call led Helen to cast aside her books for something more important than ambition: a sick child. In response to Mary's panicked call that no one knew what was wrong with her violently ill toddler, Helen won her professors' agreement to postpone her exams and boarded a train to Boston. There she tended to her niece, Polly (Mary "Polly" Henderson), who was diagnosed with a hernia, until her recovery was certain.

Whether Helen's grades were posted along with those of her classmates is doubtful, since they might have shown that some women locked out of slots because of their gender had scored higher than the men who won them. But soon enough she knew her fate. Vivian Tappan had beaten her by two-tenths of a point. Tappan, not Taussig, would get the one slot available for a woman in internal medicine, the pathway to become a physician of the heart. Now what would Helen do?

4   LITTLE CHOICE

Boston was where Helen ultimately hoped to work, but even as she plotted her return after the disappointing loss of the residency, she felt fortunate to obtain a short-term research fellowship for further graduate study under one of her favorite teachers, Edward P. Carter, who ran the adult heart station at Hopkins. Friends had suggested she apply for a pediatrics fellowship, but Helen deemed the prospect too risky; the leadership of the pediatrics department was in limbo following the death of its founder, John Howland. Howland did not approve of women doctors and had even fired his chief nurse, an outspoken suffragette. Margaret Handy, who ran the outpatient clinic in 1918 when half the faculty was overseas, had been hired only after pledging not to get married for forty years.[1] Helen, a self-described upstart, believed he would never have hired her, and she was wary of whoever might take his place. Her attitude toward pediatrics changed, however, when, at Carter's request, she volunteered on Saturdays at a new children's heart clinic, the pet project of the new pediatrics chief. His name was Edwards M. Park.

Dr. "Paahk" as Helen pronounced his name was 6'4", with a beguiling smile and round steel-framed glasses. On many Saturdays, Park himself showed up at the clinic and stayed the afternoon. He was down-to-earth, whimsical, and warm; regularly he called home to his wife, Agnes, to give her a head count of pediatric residents for dinner. He had a history of hiring women. Very quickly he and Helen struck up a friendship that would unfold amid the backdrop of the specialization of medicine.

Helen would soon find herself the leading player in Park's struggle to upend the department of pediatrics and its model doctor-training program.

He had in mind a change that switched the focus from training doctors to treating patients. He believed both would benefit. The first indication of Park's intent to focus on the welfare of children was his decision in 1928 to arrange their hospital beds by contagious and noncontagious conditions rather than by the color of their skin, a bold move in a hospital where even in the morgue, babies were segregated by race.

*

The science of pediatrics—named for the Greek words for healer and children—was very much in its infancy in 1927. Doctors had realized during the 1800s that they needed to treat growing children differently from adults, and pediatrics began in earnest in 1894 when L. Emmett Holt published his best-selling book, *The Care and Feeding of Children*. Holt ran his own laboratory inside Babies Hospital in New York City where he employed chemical tests on children's blood and urine to understand their development processes and discover and treat their ailments. Other cities also had hospitals for children—notably Boston and Philadelphia—and some doctors in private practice began devoting themselves to children.

Medical schools did not much bother with children, relying on practicing pediatricians to teach in their spare time. No medical school had full-time faculty or a hospital devoted to children until 1912 when Johns Hopkins opened a gleaming new children's hospital on the campus and hired Holt's assistant at Babies Hospital, John Howland, to oversee it. Howland hired a handful of pediatricians including Park, who had worked with Holt at the Columbia University College of Physicians. The salary freed Park and the others to research and teach and still spend half their time treating patients.

The five-story Greek-columned building had multiple waiting and exam rooms, hospital beds on upper floors, classrooms and laboratories on each floor, and a free outpatient clinic. For the first time, faculty and students alike could examine children in a clinic, investigate their illnesses using biochemical techniques in the laboratory, and follow them in the hospital. Medical students could learn to identify childhood infectious diseases they had only read about. This hospital-based teaching laboratory was copied by other medical schools and led to the system of medical residents, or doctors in training.

Officially the hospital was called the Harriet Lane Home for Invalid Children, named for its benefactor, the most famous woman in her day, a blond, blue-eyed charmer who served as first lady to her unmarried uncle, President James Buchanan, and afterward married into a wealthy Baltimore family. Harriet Lane lost her two sons, Buck at fifteen, and Henry at twelve, to rheumatic fever. She and her husband, Henry E. Johnston, intended to use their wealth to build a convalescent home for rheumatic children who suffered for years with the disease, but the $400,000 remaining when she died in 1904 was not enough. Instead, aware Hopkins doctors wanted but could not afford a children's hospital, her trustees agreed to erect and maintain the building if doctors assumed responsibility to run it.

Park ran the free outpatient clinic. Under his direction, students learned to set a broken limb and hydrate a child. They also learned to identify contagious but curable diseases like measles, mumps, whooping cough, and diphtheria under a microscope in the laboratory. A child with a broken leg went home in a cast. Children diagnosed with contagious diseases recovered at a city hospital. But for children with recurring or chronic diseases that were not contagious, diseases like rheumatic fever, which attacked and weakened the heart, Park saw that the model failed. Patients were treated for flare-ups of their symptoms, sent home, and soon forgotten. Many of these patients died in childhood or early adulthood as the disease, unimpeded, ravaged their bodies.

The idea of observing children suffering from chronic conditions to prevent premature death emerged in the 1920s among leading researchers at urban hospitals, including Park's Columbia medical school classmate, Arthur E. Cohn, now at the Rockefeller Institute for Medical Research and Hospital.[2] It was fed by advances in scientific techniques and technology, like the electrocardiograph (EKG) Cohn first brought to America in 1909 from England, that allowed detailed assessments and greater efficiency. Medicine was embarking on a variation of the scientific management movement that Helen had heard discussed a decade earlier at her dining room table in Cambridge.

Park demonstrated the success of the concept himself, as Helen would learn. Under Howland at Hopkins, he had studied rickets in the laboratory. But at Yale, where he was hired to open a pediatrics department, Park and a team including Martha Eliot, the Hopkins-trained doctor, began carefully observing children with bowed legs, easily broken limbs, or other deformities linked to soft bones—symptoms of rickets. The more cases of rickets they saw, the better they got at identifying the progression of the disease in the bones and, ultimately, what caused it: a lack of vitamin D to help the body absorb calcium and keep bones strong. Treatment was obvious as a result: cod liver oil, a food packed with vitamin D.

In Baltimore in 1928, Park was bent on opening clinics organized around problems of the body: the heart in rheumatic fever patients, the lungs in children with tuberculosis, the brain in children with psychiatric problems, and the kidney in children with diabetes. The Saturday heart clinic was a trial run of his strategy to marry laboratory science to sustained patient care. He faced an uphill battle. Pulling children from the general outpatient clinic to be treated instead in the new specialty clinics met immediate pro-test from the house staff in pediatrics. Of all the specialty clinics, none was more unpopular than the heart clinic. If residents had to send children with murmurs to the heart clinic, how could they learn to diagnose rheumatic fever? The clinic even closed temporarily while Park searched for money to hire a social worker. He believed doctors needed to understand chroni-cally ill patients in their home settings to treat them. He got it from a new private foundation focused on social needs including childhood health, the Commonwealth Fund. His friend Cohn at the Rockefeller Hospital was designing tests to study heart conditions and influenced which projects got funded.

Helen knew enough about Park to want to learn from him. She applied for an internship in general pediatrics starting in fall 1928. Throughout her internship, she continued to volunteer in the children's heart clinic, which was now open three days a week under the temporary direction of a pediatrics resident, Dr. Clifford Leech. It was the biggest of Park's clinics and offered the most potential for treatment advances. But pediatric residents viewed the work as boring, even useless, since children inevitably died—an endless grind of data collection, papers on birth defects, and autopsy reviews. A woman had fewer opportunities and might stay in the job longer than a man, Park reasoned, and Helen knew more about the heart than any of her classmates. After watching Helen examine patients, Park decided he wanted her to head his new children's heart clinic. In January 1929, only four months into her general pediatrics fellowship, he offered her the job.

Helen was in New York for six weeks to study the treatment of conta-gious diseases at the Willard Parker Hospital, the largest public hospital in the country where patients with such diseases were isolated. After months of treating sick children, Helen herself was ill with whooping cough. Park's concern was evident. Helen's health was the first subject he raised when he wrote Helen a few days after her departure. "Are the sky scrapers [*sic*] of New York City shaking with your whoops? Won't you write me in regard to your health and in regard to the value of the Willard Parker to you?"[3] The job offer came in what was very nearly a postscript: Would Helen like to direct his new heart clinic for children? He proposed that she succeed Leech beginning July 1 as the permanent head of the new heart clinic for children. By return mail, Helen refused. She needed more hospital experience, and her experience dealing with acute cardiac conditions in children was virtually nil, she wrote: "Why should my classmates especially value my opinion?"[4]

It was not the only reason she turned down Park. Helen hoped to work in Boston. And she loathed filling out the charts the Commonwealth Fund required in exchange for its financial backing. Helen thought the data were useless since what she considered a palpitation another doctor did not. She had even traveled to New York to discuss the matter with Cohn. How could she head the clinic if she was not convinced of the value of these records? Park disagreed with Helen's assessment of the charts' value and told her so. They were part of the data collection he and Cohn hoped would shed light on changes in the heart caused by disease. In any case, Helen told Park, "I must confess a tiny problem of one's own is much more fun than collecting statistics for others." Appreciative nonetheless, she affirmed her interest in the heart and agreed to discuss the possibility when she returned.

Apart from reassuring Park that New York City skyscrapers were so solid she had given up trying to shake them, Helen avoided an assessment of her cough. The truth was, the infection had been with her so long that it had moved into her middle ear, a sign that it could do permanent damage. When the coughing finally ended and she returned to Baltimore, Helen realized that her ability to hear in one ear had dropped abruptly. Deadly in children, whooping cough, or pertussis, left so many adults with significant hearing loss in the days before vaccines that public schools routinely offered night courses on reading lips. There were some doctors with poor hearing, but none who intended to make her career in a medical specialty dependent on listening. A doctor listened to sounds inside the heart made by blood swooshing through its vessels and chambers to compare them with the sound of a healthy heart. Sounds were clues that revealed a problem or disease or led a doctor to suspect an unusual event inside the body.

A development in the children's heart clinic gave Helen the reprieve she needed to adjust to her hearing problem and gain more experience treating children during the remaining eight months of her fellowship: the doctor running the heart clinic decided to remain for a second year. In hope of finding a permanent job in Boston, Helen secured a special internship and research opportunity in the pathology lab of T. B. Mallory at Harvard. When her stepmother died in January 1930, Helen assured her father that she believed she had found a good enough position beginning February 1 that she was willing to quit Hopkins and move in with him.[5]

Park continued to woo Helen. He wrote as soon as she left, reminding Helen of the job awaiting her in Baltimore. In Mallory's lab, she learned new techniques to examine the adult heart, and on her own time, she followed along on hospital rounds in general pediatrics twice a week, relying on connections to pediatricians trained at Hopkins. They also invited her to the weekly children's cardiology clinic and introduced her to doctors specializing in the heart. They included Paul Dudley White, a Harvard cardiologist who was to link lifestyle to heart disease in adults; T. Duckett Jones, who would establish the criteria to diagnose rheumatic fever in children; and a visiting celebrity physician from Canada, Maude Abbott. Abbott was making a career of collecting autopsy specimens of abnormal hearts from around the world.

"Last night there was an interesting informal meeting in Dr. Paul White's laboratory in which Dr. Maude Abbott showed a most extraordinary heart," Helen wrote to Park.[6] It was the first of her many meetings with the legendary Abbott.

Helen's activities in Boston made Park even keener to hire her. He admitted to a little jealousy, saying the Bostonians were slightly more advanced in their techniques. She in turn wrote of her disappointment at being unable to produce original research for Mallory with the techniques she had learned in his laboratory. In April, just when her research looked promising, she injured her right arm and could not work.[7] And her hearing had not improved. Sometime that winter or in the early spring of 1930, the hearing in her second ear also dropped abruptly.

Uncertainty prevailed. On April 3, before leaving with her father to open their house in Cotuit, she inquired how to obtain a license to practice medicine in Massachusetts. At the same time, she told Park she was leaning toward the job in Baltimore.

Park was more excited about Helen returning to lead his clinic than any other development in his department, he told her. "I have the greatest confidence in you and believe you are your father's daughter," he wrote. In a letter of April 12, 1930, Park offered Helen an entry-level instructor's position at a salary of $1,500, assurance that the clinic's financing was secure, and freedom to choose her own research topics. He had even replaced the Commonwealth grant so Helen no longer had to compile data. "I shall let you organize things in any way that seems best," he said. He also pledged to provide her "whatever facilities I can."[8] But in her reply four days later, Helen asked for time to consider the offer. Neither Helen nor her father believed she should sacrifice her life for him, she told Park. But she wanted to ensure that nothing available in Boston was as good as the opportunity in Baltimore. When a position in Duckett Jones's pathology laboratory provided no opportunity to treat patients, Helen turned it down. On April 29, she wrote to Park. "The dye [sic] was cast last night in favor of Baltimore. . . . You know, without my telling you, that professionally I am tremendously glad to come back to your department. . . . I hope and shall try hard to make the most of my opportunities and not disappoint you."[9]

It was a gamble for both. Park had put his new children's heart clinic and his hope for a breakthrough in rheumatic fever in the hands of a woman with little experience treating children, or anyone else for that matter, and one with imperfect hearing. Helen, for her part, agreed to take charge of a clinic that her peers had shunned. She would have to convince them to send her patients. She could do little for the sickest ones. Her classmates warned against it. Treating adult patients with heart disease was one thing, but to limit herself to children?[10]

Freedom to choose her research projects was key. Park offered ideas, in the form of interesting questions. He had never been satisfied by pathologists' explanations of why the heart of a diphtheria patient grew inflamed or why it suddenly failed in patients with paralyzed throat muscles, he told Helen. He had just made a deal to allow Hopkins staff to work at Sydenham, the city's long-term care hospital for children named for the great British physician Sir Thomas Sydenham, who first described infectious diseases. Doctors treating rheumatic fever seldom had the opportunity to see heart damage caused by other diseases.[11]

The humble tone in which Helen had considered Park's job offer soon gave way to a firm, take-charge voice, over the purchase of a portable EKG machine to record the heart's electrical activity. The EKG allowed doctors to monitor changes and draw conclusions about damage to the heart muscle that a stethoscope could not pick up, and to treat a patient more rapidly. It led doctors to realize that the heart actually changed in rheumatic fever patients. Park had promised a Victor salesman he would buy one in the fall but asked Helen to investigate the Sanborn, by a Boston manufacturer. She argued for the Sanborn. Her multiple letters describing demonstrations of the machines in New York and Boston, her consultations with experts, pros and cons of specifications, and her recommended month-long test of both if Park disagreed so overwhelmed Park that he told her to decide.

The Boston manufacturer was one of a handful of companies that also produced hearing aids. Could there be little doubt but that Helen, already in the habit of consulting experts, spent at least part of this crucial summer before her new job investigating devices to improve her hearing?

*

Listening to the body to pinpoint ailments dates to 1761 when Austrian Leopold Auenbrugger put his ear on a patient's chest and tapped it lightly with his fingers, hoping to learn from the sound whether the lungs were clear or filled with fluid. Physical examinations—observing, touching with the hand to feel organs or vibrations below the surface (palpation), and tapping the chest with the fingers to feel vibrations (percussion)—to differentiate and describe heart conditions became popular after 1808 when French heart specialist Jean Corvisart translated Auenbrugger's work and expanded on his techniques. Ten years later, Frenchman René Laennec designed a hollow wooden tube with a bell on one end—the original stethoscope—to better detect sounds of diseased hearts. It reduced the need for doctors to touch the patient directly. In 1930, the best way to listen to the heart was through a stethoscope. This Y-shaped instrument, with its long-stemmed arms connected to a bell-shaped open receiver placed on the chest, allowed a doctor to listen with both ears. It improved though did not amplify sound. A stethoscope that relied on vacuum tubes to magnify sound had just been unveiled. The electronic stethoscope was still to come.

In fall 1930, when Helen left her father in Boston and returned to Baltimore to take charge of the children's heart clinic, she was already aware that the stethoscope would not be her instrument. She had begun to investigate alternatives.[12]

*

Helen settled into a rented apartment in a house under the shady elms of Lake Avenue abutting the city's most prestigious new neighborhoods. While she arranged her home, Park told Helen about a piece of equipment he ordered for the pediatrics department that, along with the stethoscope and portable EKG, had great potential to aid a doctor's diagnosis. It was called a fluoroscope, a type of X-ray that allowed the doctor to project continuous energy (X-ray beam) onto a body and illuminate the heart. The doctor could stand behind the patient and watch the beating heart on a screen, in real time, and photograph it from three angles. Park had read studies at New York Hospital by the physician May G. Wilson. Using this tool repeatedly on rheumatic fever patients as their disease progressed, she had discovered that the left chamber of the heart, which pumps oxygenated blood to the body, enlarged as rheumatic fever developed. This information about the disease might help contain it.

Helen was not initially impressed by this visual enhancement of the heart's workings, although, at Park's urging, she consulted Cohn on the merits of an attachment to the fluoroscope to produce the most accurate photographs. Instead, amid her growing deafness, Helen worried over how she would be able to distinguish heart conditions indicated by subtle sounds. She focused on a different instrument: her hands. That first year, she worked to enhance her ability to hear the workings of a heart as it was done in the 1800s, by placing sensitive fingers on a child's chest to detect sounds produced by the anatomy of his heart. Every medical student learned a cursory version of feeling vibrations on the skin to try to identify problems in the heart, but Helen, unable to count on the stethoscope, set out to master the art.

The heart's normal sounds or vibrations are caused by the closure of valves as blood flows into the heart's upper chambers or rooms and then into the two bottom chambers, the ventricles, which push it through the great arteries (pulmonary and aorta) to the lungs and the body at large. By palpating the pulse, Helen could detect blood pressure almost as accurately as doctors using the stethoscope to listen to the heart. She could also obtain the heart rate and rhythm by feeling the pulse.

Then, with her hands on the chest, Helen learned to recognize extra or unusual vibrations—murmurs—that told her something was wrong.

Abnormal sounds are caused by turbulent blood flow, for instance, when blood pushes through a valve opening that is too small. Doctors using the stethoscope characterized them by their frequency, their pitch, even their musical qualities—low, high, continual, or periodic, or a combination of these. They compared what they heard to the screech of a seagull or the rumble of a river. Each abnormal sound signaled a particular condition—for instance, a misshapen valve, or a valve grown stiff from inflammation, or a hole in the wall separating the two sides of the heart.

With her hands, Helen began to feel vibrations of the heart and match them to sounds that other doctors could hear through a stethoscope, to come up with the same diagnosis. She practiced listening with her hands on sick and healthy children. Eventually, when Helen thought she was good at feeling the movement of blood into and around the heart and detecting its mechanics with her hands, she sought to improve her technique by consulting experts. To find them, she visited a school for the deaf—the closest being the Maryland School for the Colored Deaf—and came away awed by what students there could feel.

At the beginning of medicine as we know it, with diagnostic technology independent of the physician on the horizon, Helen's examination of the heart gave rebirth to an old technique. Examining the heart by directly touching the chest wall had fallen out of fashion with doctors enamored of easy and accurate listening devices and skittish about closeness with patients. For Helen, the simple technique of touch would become her cardinal diagnostic instrument and also the way she met and nurtured enduring relationships with her patients.

5   LESSONS FROM CHILDREN (1930–1933)

The human heart powers two separate circuits of blood around the body to keep it alive. In perfect form, its two sides keep blood moving along pipes and into organs and tissues and cells, bringing oxygen and nutrients to the body and sending away waste in a finely tuned recycling system. A wall separates the two circuits. The right heart pumps bluish, nonoxygenated blood to the lungs, which harbor oxygen from the air we breathe. The lungs diffuse oxygen into the blood, tinting it red as it streams through. It makes its way along veins of the lungs to the left heart. The left heart, with its thicker muscle, pumps this blood throughout the body through the aorta and its branches.

The four chambers of the heart work in stages, in response to pressure. Simultaneously the top chambers (atria) fill with blood from outside veins. As pressure mounts, valves open to allow blood to flow down into the bottom chambers (ventricles). When the bottom chambers are full, the valves of the upper chambers snap shut, the first sound of the heartbeat, to prevent backflow, and valves at the bottom open, the second sound of the heartbeat, to disgorge their contents. Now blood moves out to the body or the lungs. The left chamber pushes fresh blood into the body via its main blood vessel, the aorta. The right chamber pushes blood though the other main blood vessel, the pulmonary artery, and into the lungs, where it can be refreshed with oxygen. Fill, pump. Fill, pump. In less than a minute, these pumps move a gallon of blood in the adult body—and in a baby, about a half cup.

In 1930, when Helen became director of the children's heart clinic at the Harriet Lane hospital, heart disease was the leading cause of death for

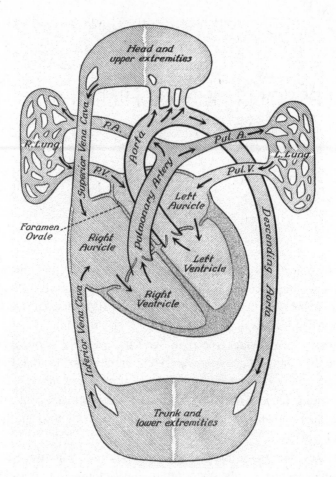

The normal heart, from *Congenital Malformations of the Heart* by Helen Brooke Taussig. Published with permission of the Commonwealth Fund.

little girls and second only to accidents for boys. The trigger was rheumatic fever. No one knew its cause or the route it might take, why some children got it and not others, why it seemed to run in some families but not others. Theories abounded over sensitivity, heredity, and environment. Wilson at New York Hospital, the doctor using the fluoroscope to capture changes in the heart, postulated that it was hereditary.[1] Doctors could not even agree on criteria to diagnose it. The only thing doctors knew in 1930 was that left unchecked, rheumatic fever attacked and weakened the heart.

As early as the 1830s in Paris, doctors using the stethoscope picked up abnormal sounds in the heart in a person with rheumatic fever.[2] In the

1920s, hospitals began routinely using the new EKG to monitor the heart's electrical system to detect changes in speed that might signal valves or muscles weakened by rheumatic fever. A weak or inflamed valve did not fully close, preventing the usual amount of blood from moving into either the body or lungs. Worse, backups caused by leaky valves could leave extra fluid in the wrong places. A heart that worked harder to push oxygenated blood through smaller valve openings became sore and inflamed. The excess fluid settled in legs and joints, causing them to swell in terrible, arthritic-like pain. Once this pileup occurred, it was difficult to reverse. Helen's treatment tools were few: she could relieve pain with a little aspirin, reduce swelling with an anti-inflammatory—a variation of aspirin—and order the only known strategy for recovery of a still-inflamed heart: bed rest. But first she needed patients.

*

Helen passed by plenty of potential patients on the way to her heart clinic, a two-story wing in the rear of the children's hospital. They crowded into the general outpatient clinic waiting room, a mass of poor mothers, sick and crying children on their laps, sitting on row after row of long, hard wooden benches. In summer they fanned themselves against the oppressive heat, and in winter, they wrapped themselves in colorful hats and cloth coats and squeezed together for warmth.

Doctors running the outpatient clinic refused to send Helen children with unusual heart sounds because they wanted to learn from them how to identify diseases of the heart. Helen could not blame them, really, since she had so little experience. When Park insisted, they sent Helen a trickle of patients in obvious, late stages of rheumatic fever, and newborns, bluish in skin tone and struggling to breathe, who would soon be dead. For the chance to help patients in the early stage, before rheumatic fever became unstoppable, Helen would have to build her own patient base. From files left by the first director of the fledgling clinic, Helen obtained the names of two hundred children with unknown diagnoses. Each day she mailed out cards to four or five families, inviting them to come in for appointments on Monday, Wednesday, and Friday until she exhausted the list. Many working-class families in the neighborhood around the hospital had lost their jobs following the 1929 stock market crash and the beginning of the Great Depression. Helen's outreach worked. Soon her appointment book filled.

*

Mysterious in its presentation and the way it played out in each child, rheumatic fever challenged the most experienced doctor. Detective work was required in the case of a child with pneumonia or flu-like symptoms since these could mask a heart inflamed by a previous episode of acute rheumatic fever. Both conditions could cause ugly and painfully swollen knees and ankles. Some children came to Helen gasping for breath. This was a clue, but to what? A rheumatic fever infection landed in a valve, and if the tender valve was on the right side of the heart, she wondered if it could stymie blood flow to the lungs. Equally plausible, the child was gasping for air because his lungs were swelled by pneumonia.

Afternoons in her clinic, Helen listened to her patients' hearts and wrote down their stories in exacting detail, noting the onset of symptoms and cataloging odd or inconsistent features. She felt the size or position of the liver, spleen, kidneys, and stomach. She studied the lines of an EKG and the image of a fuzzy X-ray. Not least was the expression on the child's face and the amount of fight in her lungs. In these early clinic years, Helen's observations from a first encounter with a patient could run ten neatly typed pages. Each follow-up visit resulted in a handwritten report of more than a page. Any detail could prove useful to someone examining a new medical theory, and not knowing which one it might be, she recorded them all. Greek, Italian, Polish, German, white, or "colored," the ethnic background and race of the child was noted in the same way she described the state of his heart and likely condition. If it was not rheumatic fever, what was it? In first-person narrative Helen unveiled her suspicions, questions, and the opinions of others she consulted and detailed her own conclusions about the processes of the heart. "[. . .] presents quite a diagnostic problem," she would begin. In the evening, at her desk at home, she analyzed her errors or those by pediatrics residents. "This was where we made our mistake," she would say.

The portable EKG machine proved extremely useful; from the pattern of lines symbolizing heartbeats left by the electrocardiogram on a piece of paper, she could detect how well a particular heart sent blood around the body or into the lungs and whether the rate, measured with each beat, was steady or stop-and-go. For children with uncertain diagnoses, Helen took a new measure with the EKG at each visit. If a heart rate had not improved, she altered treatment or changed her diagnosis. One day the EKG revealed the heart rate returned to normal in a pneumonia patient whom Helen suspected also had rheumatic fever. That ruled out rheumatic fever. "Fooled again," she wrote.

About six months after taking over the children's heart clinic, in April 1931, Helen was waiting for an elevator when Park appeared and with great excitement told her that the fluoroscope he ordered had been installed in the basement of the children's hospital that morning. Park expected Helen to document changes in the size of the heart in rheumatic patients as Wilson had demonstrated in New York. But he also wanted her to use this new capability to identify hearts that malfunctioned from birth—congenital heart defects.

Helen resisted. When she heard a damaged valve in a child grasping for air, his skin bluish from lack of oxygen, and with no obvious sign of rheumatic fever, she knew the problem was a birth defect. She examined these children and kept careful notes. But like other doctors, she had no desire to identify their specific problems. She could do nothing for babies born with malformed hearts other than ease their pain. They would die. It was work enough to identify and treat heart damage left by rheumatic fever and, when she could, to prevent the disease from progressing.

Park told Helen she had no choice: to treat children with rheumatic fever, she had to distinguish them from children with heart defects. Knowing the difference was critical, because as Helen had learned ashamedly when a two-year-old in her care died, a drug that could reduce swelling in the rheumatic child could hasten death in a child with a defective heart. But Park's command increased her workload and diluted her focus. Reluctantly, aided by a part-time secretary and occasionally an assistant resident, Helen expanded her examination of tiny babies or toddlers she knew were weeks or months from death. Henceforth, she photographed their tiny hearts from multiple angles and examined them in oblique views.

For her rheumatic patients, Helen initially found it difficult to detect small changes in the left heart with the fluoroscope. She got more reliable information with her own hands and the EKG. Borrowing Wilson's techniques, however, she quickly grasped the machine's immense diagnostic value. Compared with X-rays, which captured a single still-life angle, the fluoroscope offered a movie of the heart. By watching it work, Helen could be certain that a lung was inflamed by disease and not merely the result of a baby crying. Tilting the child from side to side, she could see the

enlarged left hearts of children with rheumatic fever and assumed they sustained valve damage, damage that she hoped was temporary. For children
who had struggled to breathe from birth, the ones with bluish skin, Helen
noticed they all seemed to have inflamed right hearts (ventricles), evidence
of how hard the right heart was working to pump blood to the lungs for
oxygen. Something was in the way. The path to the lungs, the pulmonary
artery, was either obstructed or not working at all. This was one of the first
clues Helen used to diagnose and analyze defective hearts.

Less than a year into her new job, Helen was working twelve and fourteen hours a day. To live in the style she was accustomed, she needed help.
The woman she hired in 1931 to manage her house remained in her employ
for more than three decades. Amy Brown Clark was a year older than Helen,
married, childless, and a lay leader in the Christian ministry of Sharp Street
Memorial United Methodist Church, the city's most prominent Methodist
church for "coloreds" and its foremost advocate for civil rights. In return for
cleaning, shopping, cooking, serving dinners Helen hosted after work and
on Sundays, and picking up Helen at the train or plane, Amy got a decent
wage, interesting conversation, and, presumably, superior health care. She
would be described as Helen's "faithful friend and loyal companion" as
well as her employee.[3] Fortunately for Helen, whose cooking was limited to
a few dishes she learned in Cotuit, Amy was an excellent cook. She shared
Helen's knack for flower arranging. Little more is known of this woman
who helped make it possible for Helen to succeed in the same way wives in
this era supported their spouses. But over the years, Helen would say, "We
became devoted friends."[4]

*

A photo of Helen from the early 1930s shows a clear-skinned, strong-eyed,
confident woman, her hair parted in the middle. Patient records from this
time reveal her searching voice and growing confidence, as well as her frustration, sorrow, and hope. "It is disheartening to see so much develop in
so short a time," she wrote of one little girl. "Would that we could keep
her well," she wrote of another. Helen's patient records also reveal how she
handled a worsening health problem of her own.

Once Helen began treating a rheumatic child, she listened for changes
in the sound of the heart to track the disease's progression and adjust treatment. For months and often years, she needed to remember from visit

to visit how each child sounded and how one child's sound compared to sounds in other children with the same disease at a similar point in time. The sounds of hearts damaged from birth also changed as the child grew and her body adapted to the defect. Some of them mimicked a heart damaged by rheumatic fever disease. Between 1931 and 1933, after Park required Helen to study hearts damaged at birth, the number of sounds Helen needed to master to diagnose and treat children grew exponentially. These were the very years that her hearing worsened. By 1933, Helen had become, in her own words, "quite deaf."

With her fingers or clenched palm—she called it a "claw-hand"—Helen could easily detect heart murmurs other doctors heard with the stethoscope. The sounds she found hardest to detect with her hands were also hard to detect by doctors using stethoscopes. These were the faint diastolic murmurs—unusual sounds as the heart chamber filled up with blood indicating either reverse blood flow or a partly closed valve—and pericardial friction rubs—the sound of two walls rubbing against each other, indicating a loss of fluid or lubricant in the sac around the heart.[5] This sound could mean inflammation, perhaps the result of an infection. To improve her ability to distinguish sounds, she practiced techniques she learned from deaf students. In her living room, Helen carefully tapped the soft pillow on her sofa with her fingers. How gently could she touch and still pick up vibrations made by a concert on the radio? This was the secret to palpation of the heart. Too much pressure would make vibrations harder to detect.

Helen also asked doctors working nearby to listen to a patient when she could not be sure of a sound and recorded their opinions, with the result that she began to understand from this data when and why in the process of a disease these sounds changed; for instance, she realized that murmurs revealing a valve problem became almost unnoticeable once the heart of a rheumatic child began beating irregularly. When one boy's murmur dropped in intensity a month after Helen examined him in spring 1933, she wondered whether his inflamed aortic valve had healed or whether the sound of this condition varied. She reassured herself she was not the only one confused. In the patient's file, she quoted Sir Thomas Lewis of London—the top cardiologist at that time—that no injury caused so much difference of opinion as a weakened aortic valve (aortic insufficiency). This critical valve opened to allow the left heart to push blood into the body.

As she grew confident using her fingers to detect sound, Helen challenged diagnoses made by her mentor. Detecting sounds of a weakened valve (mitral valve insufficiency), in which blood leaked back into the upper chamber instead of continuing to the body, Helen diagnosed rheumatic fever in a six-year-old with unusual swelling in her fingers. Park, seeing changes in the lungs, diagnosed tuberculosis. A guinea pig injected with the child's lung fluid after she died showed no sign of TB. Helen was right.

When she diagnosed a heart defect from birth, Helen checked it against what she found in an autopsy. In spring 1933, just when Helen thought she had mastered the sound of one of the most common heart defects in a newborn, a patent ductus arteriosus, the results of an autopsy of a baby left her discouraged.[6] In the fetus, a small passageway, the ductus arteriosus, develops between the aorta and pulmonary arteries to take oxygenated blood from the placenta around the fetus's nonworking lungs and into the body. This bypass closes on its own after birth once the baby begins using his lungs. If it remains open, or patent, the duct siphons some of the oxygenated blood going to the body via the aorta (which has higher pressure) and sends it to the infant's lungs, sometimes overwhelming them. It makes a distinct sound, a continuous murmur, and is easy to detect with the hands.

The newborn Helen examined had trouble breathing and would not eat. From the sound on the baby's chest, Helen theorized that the baby suffered from a hole in his heart wall (septal defect) separating the right and left bottom chambers. It allowed blood destined for the lungs to mix with oxygenated blood making its way to the body, reducing oxygen content. An autopsy showed the heart wall was solid. Instead, the boy had an open, or patent, ductus arteriosus.

The case showed Helen how difficult it could be to distinguish between these two birth defects in the first months of life. Why was she unable to hear the patent ductus arteriosus in some babies while in others it was marked? In the future, when she examined a child with unusual sounds, she decided to move her hands farther down the chest to be sure to detect a still-working fetal duct. This refinement in where she placed her hands would help Helen discover that in babies with defective hearts, this little duct often plays a big role.

Emboldened, Helen startled doctors in the outpatient clinic one day by suggesting that a three-week-old with troubled breathing, blue around the nose, and loud noises coming from his chest had his major blood vessels attached to the wrong sides of his heart. Helen knew of such cases from Abbott's descriptions. "Certainly there is nothing we can do for this congenital malformation," she wrote in the baby's record, "and we can but watch the course of events."

*

A correct diagnosis was only the first step in Helen's struggle to save a child with rheumatic fever. The hearts of rheumatic patients remained inflamed months and sometimes years after pain relievers kicked in and the joint swelling common in the acute or symptomatic phase had passed. She experimented with codeine and varied doses of anti-inflammatories to children in this next active phase of the disease to calm their enlarged hearts. She would be among the first to recognize that aspirin reduced inflammation, as measured by the rate red blood cells settle at the bottom of a test tube (sedimentation rate).[7] As a result, as early as 1934, she began regularly testing asymptomatic children long after the withdrawal of the drug for signs of an active infection. She also experimented with treatment children dreaded above all, bed rest, to try to prevent irreversible scarring of the heart valves. The importance of bed rest in healing the heart had been demonstrated by one of the earliest patients in the heart clinic during Helen's fellowship days.

Dottie Worthington was five when she experienced her first attack of rheumatic fever.[8] It revealed itself in swollen, bright red ankles. She recovered after spending time under an oxygen tent. Several years later, her parents noticed a shortness of breath and brought her back to the cardiac clinic. Alarmed at this relapse, Helen's predecessor, Leech, sent the girl to the recently opened St. Gabriel's Convalescent Home, run by an Anglican order of sisters, where Helen met her. For six months, through Christmas, Dottie stayed in bed under the nuns' care. To the girl's relief, she grew well enough to return to school and participate in activities without restriction. But beds in convalescent homes were scarce. And parents caring for six, eight, even eleven children balked at bed rest for a child who seemed outwardly well. When Helen prescribed months of bed rest for a ten-year-old girl whose sibling had died from the disease, the girl's mother called the idea "absurd." She recorded the mother's parting words: "Well, we all have to die sometime."[9]

In search of ways to balance a child's needs against the cost to the family, Helen visited their homes, asked questions about siblings' health, and scoured reports from her social worker for useful information. The lenient schedule she devised for one nine-year-old girl with "trouble brewing all summer" in her heart—bed rest from 6:00 p.m. to 10:00 a.m. and two hours in the afternoon—might have worked except the girl missed her next appointment. Two months later, she showed signs of Sydenham chorea, also known as St. Vitus's dance, a neurological complication in which children exhibit uncontrollable movements of their arms and legs, first identified in 1686 by the British physician Sydenham. Helen blamed herself for the girl's decline and determined to have the courage of her convictions to treat a child "promptly and drastically."[10] Thereafter, she called each patient into her private office to "firmly" recite her experience and reasons for assigning bed rest and "just as firmly" let the child's mother know the consequences of failing to provide it. Her stern, brisk manner when she had to assign him bed rest repeatedly over four years so frightened Raymond Gebhardt that he could recall it forty years later, but from the twinkle in Helen's eye when she smiled and laughed, the small boy also sensed her kindness.[11]

To get families to cooperate, Helen also developed a network of scouts. It included nurses from the Babies' Milk Fund, which provided prenatal care and free milk to poor mothers, visiting nurses, and her own social worker, Mildred White. They checked that beds were clean and children in them. They told her when children grew sicker. Helen sent an ambulance to pick up one seriously ill boy after she learned from a visiting nurse that the boy's mother had refused to bring him in. She could not round up an eleven-year-old patient Miss White spotted on the street with an obvious form of chorea. Within a decade, the boy was dead.

One day when Helen herself was forced to send home an acutely ill toddler because hospital and convalescent beds were full, she handed his mother a restraining jacket. She also freely prescribed codeine, so that the child slept in the morning, afternoon, and through the night. On one home visit, seeing a three-year-old "colored" boy who would not stay still, Helen drugged the boy and, with the help of Miss White, tied him to his bed. (Physical restraint in hospitals was routine.) Helen told the mother to continue the medicines—codeine and, for joint pain, a five-grain tablet of Pyramidon, a new drug that doctors in New York in 1930 reported superior to salicylates—until she could get a hospital bed. Despite a history of rheumatic fever, twenty years later his heart showed itself no worse for wear.

Eventually, based on her tests and careful observations of the disease, Helen would decide that a recovering child should not be allowed even to sit up in bed until two weeks after he was off medicine and free of all symptoms of active infection, even a slightly raised temperature, and any irregularity in pulse and respiration.[12] Well-off children in this era spent their formative years in bed, cared for by their mothers and drop-in doctors; a generation of famous writers attributed their careers to reading books during a sickly childhood. They were the lucky ones. Poor children lived in crowded apartments and if they got a bed in a convalescent home, they could go months without a parental visit, sometimes until Miss White gave the family taxi fare.

*

The years between 1930 and 1933 were among the most challenging in Helen's life.

She ran a start-up that required her to master a growing number of sounds at the same time she was losing her hearing. With little experience or help, she was trying to save children from a disease she could not easily identify and for which there was no known cause or standard treatment. For children she knew she could not save, she experimented with old drugs like digitalis and tried a last-resort diuretic, Salyrgan, a mercury compound, to reduce inflammation and help them breathe, admitting to misgivings, since prolonging their lives increased their suffering. Meanwhile, her forays into the pathology laboratory to look at malformed hearts were taking more and more of her time.

To learn about hearts that did not form correctly in the fetal stage, Helen read hundreds of pathology reports by Abbott in Montreal. In the year and a half since meeting Abbott in Boston, Helen had regularly sought her opinion. There is little doubt but that she read Abbott's exquisitely rendered chapter on congenital heart defects in Sir William Osler's book, *Modern Medicine, Its Theory and Practice* (vol. IV, 1908). Abbott collected bottled autopsy specimens of patients' hearts and categorized them to try to explain what caused their physical manifestations. When Helen considered the possibilities behind a child's condition, she referred to Abbott's descriptions as she tried to explain to herself how a defective heart might work.

Helen renewed her contact with Abbott when both were invited to present their work at a symposium of leading heart doctors in New York. In October 1931, armed with Park's letters of introduction, Helen boarded a train to New York City. Cohn, who had studied with Sir Thomas Lewis, was to introduce her to the Londoner himself. Helen also hoped for help with introductions from Robert Gross, a recent Harvard medical school graduate examining deformed hearts in the pathology lab; they had corresponded extensively that fall.

One evening, to Helen's delight, Abbott convinced Emanuel Libman, a Mount Sinai pathologist, to include Helen and a third woman doctor at a banquet for Lewis. After dinner, the men changed places so that Helen could chat with Lewis. Writing to Park the next day, Helen confided that she suspected Abbott included her so she would not be the only woman. "It was a wonderful opportunity for me to meet the leading men in New York and also Sir Thomas Lewis," she wrote. "As you have doubtless gathered, I am having a great time and I very much appreciate your letting me come and your letters of introduction."[13]

Abbott, who was visiting US hospitals that fall, stayed with Helen when she arrived in Baltimore the following month. The two women had traveled a common path. Like Helen, Abbott was denied her choice of medical school because of her gender. She too became captivated by the heart and had a mentor who recognized her talent. While Abbott examined lifeless organs in the lab and theorized how they worked, Helen was doing science both in the laboratory and at the bedside, assessing a child's symptoms with a variety of measures, and developing her own understanding of how the heart worked. She wanted to identify defective hearts in a living patient. Abbott was happy to help. The gregarious and fun-loving doctor routinely kept Helen up nearly all night talking so that Helen, who ordinarily retired early, was useless the next day. Instead of staying with Abbott the next time Helen was in the same city, she found a room elsewhere.

*

While Helen was diverted by malformed hearts, one of the first rheumatic fever cases treated in the clinic had come to haunt her. Dottie Worthington, who had led a normal life for three years after she had been nursed to good health at St. Gabriel's, exemplifying the benefits of bed rest, had taken a turn for the worse. Earlier that year (1933), Helen treated Dottie, now in middle school, for an infection caused by breathing difficulties. After a week's rest at home, she returned to the clinic and seemed fine. Another doctor told her to return in a month, instead of a week, which would be two weeks from the acute respiratory infection and exactly when Helen knew to expect a flare-up of acute rheumatic fever. By the third week, Dottie could no longer walk. By May, her legs were three times their normal size.

On a visit to her home, Helen found the child hardly breathing, half propped up on bed on her right side and leaning forward to catch the air by allowing her heart to drop forward. Helen heard a churning murmur indicating her heart was big and her liver so engorged it was about to erupt. Helen did not think she would walk again. If Dottie survived this attack, she would never survive another. Gloom prevailed. Helen suggested Salyrgan, which had worked so miraculously for her other patient, but the family's private doctor rejected it. Dottie grew so ill by the summer of 1933 that her private doctor advised he could do nothing more for her and pronounced the case hopeless.[14]

Helen took respite in London, sailing with her father and sisters to witness Frank Taussig's acceptance of an honorary degree from the University of Cambridge. The summer getaway with family was a welcome break from the sorrowful and steep mountain of daily work. Helen stayed with her sister Catherine in Oxford part of the time, and in a pocket-sized travel diary she recorded scenes from the life other women of her social class regularly enjoyed: visits to museums with family members, a game of tennis, and high tea every afternoon. The family investments in railroads and utilities had insulated the Taussigs from the Great Depression, as did the stewardship of Frank and his relatives in St. Louis. Near to their ancestral homeland, German chancellor Adolf Hitler had taken power, quashed democratic institutions, and opened a boycott of Jewish businesses. Frank and his family, while not Jewish themselves, maintained relationships with family members who were Jewish. Beginning in 1933, Frank received letters from refugees weekly, sometimes twice a week, that appalled him. One distant relative who sought his help had lost his job with a German oil company in Nuremberg because he was Jewish and fled to Rome. Frank, aggrieved, recommended him for a job with Standard Oil.[15]

Helen took advantage of professional opportunities too, organizing or attending lunches and dinners, sometimes taking her father. With an introduction from Park, she visited doctors at London hospitals. On a tour of wards for rheumatic fever victims, she saw a bed rest regiment even stricter than hers. With Lewis one morning, she discussed clinical questions and watched him conduct patient tests. "That was a treat, and a profitable one," she told Park.[16]

In Baltimore in late September, she resumed her frenetic pace. By November, with no one to relieve her, Helen decided against joining her father, brother, and sisters in Boston for Thanksgiving. Visits to Boston and Cotuit had grown further apart.

When Helen could give no more to her patients, though, they surprised her, gave her something back, as if refreshing her so she could cross further into the desert. On the afternoon of November 10, 1933, the grand entrance of one of her sickest patients shocked Helen. Dottie, the little girl given up for hopeless by her private doctor, walked into her clinic unaided, accompanied by her mother. She "was beaming all over and looking like a million dollars," Helen recounted in a report she wrote that evening. Dottie's heart was larger than it had been during Helen's spring home visit, and in her exam, Helen heard a harsh (systolic) murmur (a possible mitral valve infection) and a definite suggestion of an abnormal heart (gallop) rhythm. Her liver was still enlarged. Things had by no means quieted down, but she was walking and back in school, wisely or not, Helen thought.

Even more remarkable than this stunning reversal were the events that led up to it. After the girl's doctor gave up, a relative persuaded her to try a lay healer. The child tried it "faithfully and hard," Helen reported. Initially, the healer read with Dottie in her home for hours every day. Then the child went to church to read. Between the end of summer and the November day she walked into Helen's clinic, she had had only one bad breathing spell.

Emotion flew from Helen's pen as she recounted these events. She began by chastising herself grandly. She had made a terrible mistake, she wrote. The mistake was in thinking the child would never get well and, worse, allowing the child to realize that doctors didn't expect her to get better. For nearly three months, the child had lain desperately ill, visited by no one who thought she would ever get well or that they could ever do anything for her. "If one is desperately and critically ill," Helen wrote, "it must be frightfully discouraging if everyone around thinks that you are going to die."

The great thing the healer did was to convince the girl to believe she would get better, Helen wrote. "I have always thought that the will to live and the will to die were very strong and very real, and I cannot but think that in this case it made the difference between life and death," she wrote. "The case proves to me how very wrong it is to surround a patient with a hopeless atmosphere no matter how desperately ill one may think she is. I hope I'll never forget this child and will have the courage to cheer other children who are desperately ill; not that I think it is the only factor, but that it is a very real factor."[17] The lesson would be woven into some of Helen's published papers as well as into her private discussions with parents. The case would change the way Helen practiced medicine, tying her even closer spiritually to her patients; hope was something real, a gift, and it was linked to a person's will. Hope was the mystery that propelled one to find new facts, new ways of doing things. The key ingredient in an Emersonian recipe for living that Helen extolled: the "romantic expectations" Emerson wrote that, when linked to a life of learning, led to advances in science, art, and poetry.[18] Who knew what would happen when a person had a goal and "tried it faithfully and hard"?

6 BREAKDOWN AND BREAKTHROUGH

Every morning on her way to her office in the hospital, Helen passed a marble plaque honoring Harriet Lane, the mother who hoped to prevent more childhood deaths from rheumatic disease by building a children's convalescent home. Other than newer anti-inflammatory drugs to slow the disease, which Helen used to advantage, not much had changed in the fifty years since Lane's sons had died. The disease was so cruel that Helen found signs of heart failure in a two-year-old boy that normally developed over years. She administered oxygen to a girl blue all over from respiratory trouble, with a heart so large it protruded from her chest, and sixty-five nodes all over her—Helen counted them—but the girl died the next day.

If winter was relentless, spring 1934 spiraled downward. The parents of Dottie Worthington, the girl with the overwhelming will to live, called to invite Helen to visit. Dottie was back in bed. Helen was on her way to a medical meeting in Atlantic City that day and declined. A few weeks later, Dottie's parents telephoned to say that the teenager had succumbed to her illness. Helen consoled them, and presumably herself; she knew how the cycle of miracle recoveries could end. Dottie lived longer than she should have and pushing back death was all that she and her parents could have expected. Later that day Helen opened the girl's file and added a simple postscript: "[. . .] died at 3:15 this a.m."

*

The business of diagnosing acute rheumatic fever and managing its chronic form, rheumatic heart disease, was all-consuming. Helen spent the first hour of her day reviewing reports and questioning parents by phone to determine if she needed to see their children at home. Then it was over to the hospital to examine acutely ill children and review their blood tests. On afternoons when the clinic was closed, she drove to patients' homes or one of two convalescent homes on the city's outskirts. Evenings, she wrote what she learned. On Saturdays, Helen might leave work earlier, in time for a hike, followed by dinner with friends or a symphony. Even with her declining hearing, she could still hear music.

Sunday mornings Helen wrote to experts about unusual cases, thanked those who helped, and tried to solicit women's groups to raise money for convalescent beds. Afternoons, though, she relaxed at the Parks' home in the countryside north of Baltimore. Helen enjoyed great conversation, and at their table, it ranged from philosophy, world events, and medical questions to poetry and art. She did not always join the others in laughter; it was difficult to follow the storyteller's lips if, upon delivering the punch line, he smiled, and she was loath to ask him to repeat it. A pediatrics fellow from Dublin who met Helen there described her as "a shy lady doctor with a diffident smile."[1]

Helen was a rare woman in a sea of men, the head of a controversial clinic with skeptics all around and hard of hearing in a field that required her to diagnose children by sound. Occasionally she bumped into Harriet Guild, whom Park recruited to head a kidney clinic for children with diabetes. Margaret Handy regularly drove from Wilmington for the day, which often ended with hikes around the Parks' property. Handy had known the Parks more than a decade, and she had advanced the practice of pediatrics, recently opening a program for premature infants. Handy was not particularly fond of Helen, though, after finding her beef hearts in the boarding-house bathtub. People sensed their rivalry. Even as women in medicine sought out each other for support, their expectations for their careers, the need to vie for scarce internships, jobs, and the attention of the rare mentor, along with differences on whether and how to meld career with a family, often led to strained relationships.

Try as she might to add to the academic literature on rheumatic heart disease, Helen could not make headway. A lack of time prevented her from publishing even minor articles on her most interesting cases. She had a paper in her head about best ways to treat rheumatics. Another paper she sent to Park in February 1933 on a six-year-old with a mysterious heart failure was in limbo until she had time to answer Park's questions. She was still perturbed about wasting time on a third paper on one hundred patients with Sydenham's chorea. Doctors called on her regularly to identify it, including to differentiate it from schizophrenia, but Park had vetoed publication after learning that a definitive study required five hundred cases. That was one paper "down the drain," she put it.[2]

Her interest in children with malformed hearts also took time from publishing. In the first few days of life, when babies were getting used to their lungs and circulation system, their bodies either found ways around defects or they died. As they developed, the odd conditions they were born with evolved. Helen tracked the changes, searching for clues. One day the death of a "miserable little undeveloped infant" led Helen to gather all the cases she could find on the patent ductus arteriosus, the temporary fetal duct that failed to close after birth. It allowed blood that was supposed to go to the body to leak into the lungs, sometimes overwhelming them.

This infant was panting, and his heart was racing at a terrific pace of 150 beats per minute when Helen first saw him. She could not see the left edge of his heart on an X-ray or with the fluoroscope. The baby went home but later returned to the hospital with a respiratory infection that proved fatal. An autopsy revealed that that he had a very large patent ductus arteriosus. Both sides of the heart had been working overtime to handle the blood that leaked from it. This accounted for a massive buildup of muscle in the heart's ventricles.

Helen became fascinated by this open duct. She researched how it worked in children with no other heart defects, from the buildup of blood in the lungs to the strain on the ventricles pushing extra blood around the body. When she reviewed other cases, she was surprised to learn that most of them showed muscle buildup in both ventricles among patients who lived with the open vessel many years. The oldest was age forty-four. She realized that babies with additional defects, such as a faulty valve or hole in the heart wall, could also live a long time with overworked hearts. "This baby taught us a great many things," she began her report.[3]

Over the next few years, Helen noticed that some blue babies lived longer if this fetal duct remained open. Their breathing problems developed slowly, sometimes not for several months after birth, when they first exhibited bluish skin. She could see with her fluoroscope that the normal path to the lungs was obscured. She could hear strange sounds. In some children she heard the murmur of a patent ductus arteriosus. It sat past the pulmonary artery and served as an alternate conduit for blood from the aorta to the lungs. As the sound grew fainter, children with narrow routes to the lungs (a flawed pulmonary valve that stymied normal blood flow) grew bluer and their breathing attacks increased. Children with a completely blocked passage to the lungs (atresia) died. During autopsies, Helen inevitably found tissue indicating that the duct had closed.

A question from a young doctor one day crystallized the problem. He had watched a baby with multiple heart defects become bluer and bluer before death, but despite an inflamed heart, the infant never went into cardiac arrest. Why had this baby died? It was not their hearts that were giving out, Helen told him. It was the patent ductus arteriosus, the duct that leaked blood to the lungs of a child whose normal path was blocked, allowing the blood to be oxygenized. When this emergency path closed, there was no other way for blood to get to the lungs. The babies were dying from a lack of oxygen.

*

By summer 1934, Baltimore stinking from decaying garbage and human sweat in the sizzling heat, Helen was unusually tired. On July 25, another teenaged girl who had fought rheumatic fever died at home. It was a difficult case. Helen had detected what she thought was a leaking valve (aortic insufficiency) to explain the enlarged heart, but the men on the medical service were skeptical. Time after time, Helen had gone to the girl's house. She improved so dramatically after Helen treated her with digitalis that Helen allowed her out of bed for an occasional automobile ride. When she relapsed, the girl refused to return to the hospital. The autopsy showed massive infection on the aortic valve Helen had suspected was leaking, but the culprit behind her death was the mitral valve. It was the valve between the two left chambers, the one attacked first by rheumatic fever and the one Helen never wanted to see damaged.

Helen left the city quickly after this. It was customary for poorly paid doctors in academe to escape to cooler climates, leaving interns and second-year medical students to care for patients. Park had already taken his family to Nova Scotia, where he lived in a cottage without electricity and fished for salmon. By the time Helen arrived after a long drive to her grandmother's house—the Taussig Big House—in Cotuit, she was in need of a long respite.

This time Cotuit could not heal her. After three weeks, she was discouraged at the thought of starting work. At the end of August, she wrote to a colleague in Baltimore, asking him to step in for her so she could delay her return. To Park's secretary, she admitted that the winter clinic had drained her.

*

When Helen returned to Baltimore in the middle of September, she tried to reimmerse herself in patients but found it difficult. Park saw the change, and the two evidently spoke. After only two weeks, Helen was on the road back to Cambridge. When her father learned this, he sent instructions to the man who managed the house in Cotuit: put away the Ping-Pong table and take out the canoe. He and Helen were coming out for the weekend.

Helen could not have found a more sympathetic soul. Frank had long feared his daughter worked too hard. Now, drawing on his own experience with mental exhaustion and that of his father, he was determined to provide Helen with rest. This is when they began talking about building Helen a cottage in Cotuit; regular getaways were more practical than the years-long hiatus Frank had taken in the Alps. But it was Helen who described the cause of her problem and proposed the solution.

From Cambridge, Helen promptly penned a letter to Park on her father's stationery. The primary problem, she told him, was that she was utterly exhausted by the year's work. "I *can not* [sic] expect to attempt another year such as last without having to pay for it again next summer," she told him. She asked of him two things: clear direction on the goal of the children's heart clinic and a limit on patients or, if the clinic was to remain large, a second full-time doctor and another social worker. Continuing a conversation evidently begun in Baltimore, Helen differentiated her job from that of the doctor in charge of the outpatient clinic. He worked quite as hard as she did, probably harder, she wrote, but only for a year. She was in her fourth year. "For the same person, year in and year out, it would be an impossible job." Moreover, a doctor in a job for only one year would not have research ideas bursting to be documented, she told him. "In any case it seems clear to me that I cannot possibly run a clinic such as mine is today and at the same time do research work or additional clinical work."[4] She asked him to decide whether the purpose of the clinic was to care for children with heart disease or to study rheumatic fever.

At stake was Park's idea that better care of chronically ill patients would lead to discoveries. He was right that research and quality patient care could coexist. Helen was learning things from long-time patients that she used to improve the care of newer patients. This included when an infection might reemerge and the timing and dose of aspirin and bed rest to combat it. But caring for rheumatic patients—often until they died—and distinguishing them from children with heart defects came at the expense of publishing what she learned and, now, of Helen's well-being.

Helen was careful to express her affection and respect for her mentor and hope that Park would appreciate her writing so freely. That this was a soul-searching time for Helen is evident not only from her decision to leave her clinic in the care of another doctor but also from the lengthy visit she made to her father's sister, Aunt Jenny, at Ladless, the Louisville farm Jenny owned with her late husband, Alfred Brandeis, a grain dealer and brother of the jurist Louis Brandeis. Helen last visited her aunt when she was in college. In need of deep rest and perhaps the advice of a seasoned woman, Helen settled in to feel sun on her skin, listen to birds and wander the lush lawns, and as she told Park, do little but "loafing, walking, playing out of doors."

After four years at the helm of the new heart clinic, Helen had published little of significance. She was thirty-six years old. Others were gaining ground. Her friend Ann G. Kuttner, who graduated from Johns Hopkins medical school three years after Helen, had begun illuminating laboratory research on repeated attacks of rheumatic fever on children in a New York convalescent home. Pediatrician Wilson with her wide roster of young patients in New York was producing study after study on its possible cause, and Duckett Jones in Boston continued to refine its symptoms in children. Caroline Bedell, who had also followed Helen in medical school, had completed an internship at Harvard, won a fellowship in physiology at Hopkins, and married that spring. Helen did not see herself as a professor in the laboratory like Bedell. She was happiest treating patients. But like her father, she felt strongly that if she learned things from her work that might be useful to other doctors, she had an obligation to share them.

Park repeatedly urged Helen not to return to Baltimore until she was wholly herself again and even enlisted Frank Taussig's help. Frank agreed to write to Helen "along the lines" Park suggested to him and expressed gratitude for Park's interest and advice, but also pressed his daughter's case for more time to conduct research. "Helen's near approach to a breakdown this summer and her slowness in recuperation have seemed to me to show that a somewhat radical reorganization of her work is called for," he told Park.[5]

Helen's question about the clinic's purpose, whether it was to treat heart patients or to study rheumatic disease, remained unanswered. Park refused to be pinned down. The only way for Helen to produce research, he told her, was for her to become "heartless" and reduce her patients to those she could care for with the available help, freeing herself to carry on her investigations.

Despite his counsel, Park scrambled to find help and money, so that Helen would not have to limit patients or change her methods to care for them. He secured part-time doctors from outside pediatrics, including from the adult heart service. He approached foundations and groups that had donated in the past. Finding them stunned by stock market losses, he opened talks with the fledgling US Public Health Service. Was the federal government interested in funding studies on cases Helen was accumulating of children with high blood pressure, or hypertension?

In the middle of December 1934, Helen returned to work. She learned she had received a small raise, which Park secured from the Harriet Lane Community Fund, a charity associated with one of the trustees of the hospital building. More important to Helen, she had written a paper worthy of being published, on managing rheumatic disease.

*

The doctor who insisted on bed rest for her rheumatic patients, who had relied on it herself in childhood to stave off tuberculosis, now embraced regular rest for mental health too. In her time away, using an inheritance from her grandmother, she bought land on Oregon Beach at Popponesset Bay within view of the Taussig Big House and began building herself a cottage in Cotuit. Frank was the sole beneficiary of his mother-in-law's estate, and he apportioned most of it among his children. Eventually each daughter would purchase a family home in Cotuit, but their children always congregated at Aunt Helen's to clam, picnic, and camp.

In spring 1935, workers finished installing wallpaper of blue and green flowers that Helen had chosen for her bedroom. Her father secured permits, purchased an extra plot of land to the west of the cottage for privacy, and arranged for a plumber. Helen traveled to Boston in April for Easter and stayed that week in her cottage in Cotuit for the first time. She returned in May for Memorial Day. With another doctor running her clinic, whom she wrote to daily, Helen spent the better part of two months in summer 1935 in Cotuit. One day in July, she stopped by the Cotuit post office for her mail and found the proof of her article on the care of rheumatic patients.[6] In the accompanying note, Park said he had ordered two hundred reprints in anticipation of demand.

*

One day in November 1935 a four-month-old baby boy was admitted to the hospital for convulsions. The baby weighed only seven pounds, and although his skin was tinted a slight blue when he was born, his mother reported that his skin turned blue only when he cried. The boy's color signaled that his blood was low on oxygen and explained his shortness of breath, which became so labored he began convulsing. By the time Helen saw him, his arms and legs were an intense blue. His color was caused by the reflection of light from the dark shade of red—nearly purple—oxygen-deprived blood in the small arteries closest to the skin.

Helen examined the boy with her hands, noting an enlarged left heart, and listened to his heart with her stethoscope, finding little else out of the ordinary. She X-rayed him and watched him from different angles under the fluoroscope. All the work of the heart appeared to be taking place on the left side, and the lower chamber on that side, the ventricle, was overly large, obviously from pumping so much. But Helen was struck by what she could *not* see. Oddly, the shadow of the heart extended out at a right angle to the breastbone. At the normal spot for a valve into the upper right chamber, the shadow faced inward instead of outward. When she turned the baby for a different view, there was no shadow at all behind the aorta, where the lower right chamber, the ventricle, should be. She concluded that the right ventricle—the chamber that pumps blood to the pulmonary artery and then to the lungs to pick up oxygen—was completely missing, and since the pulmonary artery was obscured, so was the upper chamber of the right heart.

Helen was saying that two of the four chambers of the heart, the entire right side that sent blood to the lungs to be oxygenized, were missing. If what she suggested were true, how was it possible that the baby was still alive? She theorized that blood had made it to the lungs for oxygen through the fetal bypass, the patent ductus arteriosus. If it remained open after birth, the duct siphoned some of the blood going to the body via the aorta and sent it to the infant's lungs. But no one, including Helen, could hear a murmur indicating the baby's duct was open. And her colleagues were skeptical; it was preposterous to imagine a child functioning without the part of his heart needed to pump blood to the lungs.

Helen knew of cases of missing right hearts from Maude Abbott and others and from the papers she had discovered at professional meetings or received in the mail in response to her queries. She was relieved when a later EKG showed that the heart's electrical activity was off center, a sign that the left side of the heart was doing all the work, which supported her theory. As for the absence of any sound indicating an open fetal duct, this did not bother her. She was sure that if the pressure between the two circulation systems was even, blood passing through the duct would not vibrate and could not be heard.

There was nothing Helen could do to correct this baby's heart, which she told the child's parents. But she saw herself as a healer and believed that role included offering hope. Without hope, the baby's parents had no reason to return. Helen would never know if there was something to be learned that could help the next baby. So she told the parents what she could do for their infant—small things like adjust his feeding schedule or ease his breathing with oxygen. She cared for him for a month, until he was admitted to the hospital and died there. This provided the opportunity to examine the boy's heart.

The skeptical house staff that treated the child in the hospital requested an autopsy. It was there, just as she imagined. The little bypass that Helen theorized had kept the child alive by moving blood into the pulmonary artery where it could get to the lungs for oxygen was still open. The right ventricle *was* missing, as Helen suspected. There was no sign of the wall that normally separates right and left ventricles. There was no valve connection to the missing right ventricle (tricuspid atresia) or to the pulmonary artery that in a normal baby took blood to the lungs for oxygen (pulmonary atresia). Its base was sealed shut. But the upper right chamber (atrium) Helen thought missing was hidden behind the breastbone, so the baby had three of the normal four heart chambers.

Three weeks later, doctors in the hospital asked Helen to examine another extremely ill baby, who was breathing noisily and oozing blood from her nose when Helen stopped at her crib in the hospital ward. A staff doctor threw an X-ray of the child up on the window to review it, then apologized and began to remove it. He assumed it was the X-ray of the baby boy with the missing right ventricle. "No, it isn't," Helen stopped him. "It is another child with the same malformation."[7] This little girl also had been blue at birth, but aside from a runny nose, she had improved considerably in the first ten days. She was a well-nourished eight pounds, though her mother, upon reflection, told Helen that from the start, the baby's fingertips were blue. At four weeks old, doctors treated her for bruises on her thighs. At five weeks, both of her hips began to swell, and her parents brought her back to the hospital.

Under the fluoroscope, the shadow of the heart appeared identical to that of the first baby. This time Helen was sure that the infant girl was missing two of the four chambers of the heart, on the side that pumped blood to the lungs. Helen again theorized that the baby was alive because the little duct normally used only in gestation was still operating. This extra passageway between the pulmonary artery and the aorta allowed blood to leak into the lungs.

The baby's bruises healed, and she was released from the hospital at the end of December. Again, Helen urged the family to return to her clinic for care. On February 2, 1936, she was readmitted with pneumonia and died seventy-two hours later. The autopsy showed the baby had only two of the heart's four chambers (biloculate heart). Her right ventricle was tiny, and the passageway to the lungs, the pulmonary artery, was covered by a thick tissue and impassible. With this path to the lungs closed, the baby had to have relied on a still open fetal duct to move blood to the lungs. And there it was, still open.

A baby missing a part of her heart was rare. In the book that Maude Abbott was soon to publish, only thirteen of the one thousand heart specimens she cataloged were missing one or two of the four chambers.[8] That Helen saw two babies missing sections of heart within three weeks was even more unusual. Babies with such serious problems died at birth. Both infants lived long enough for her to see them and care for them, the baby girl more than three months. It was the first time that a doctor had been able to diagnose a specific malformation of the heart in a living person.

Until Helen examined these two babies, no one had matched conditions in a living person to what was known about a heart sitting in a jar in the laboratory. They didn't know that tiny malformations had their own consistent symptoms. Helen and the other doctors realized that there could be a pattern to heart defects and that the instruments she was using allowed them to be diagnosed. Congenital malformations suddenly began to make sense. Crucially, because of her careful observation, she began to understand how babies without adequate oxygen had stayed alive so long, how exactly their circulation worked, and why they died. This would be the first step toward fixing them.

Park was ecstatic. His vision for Helen's clinic was bearing fruit. Perhaps he wore the sparkling smile often found in photos of him when he told Helen, "Well, now, I think you've learned something!"[9] News of her discovery spread quickly. Doctors who had regarded Helen's diagnoses as odd now sent her their troubling cases. She was becoming enamored of what she called her crossword puzzles of the heart. Helen wrote to Maude Abbott, too, about her findings, because Abbott invited her along with the top names in heart research to the upcoming meeting of her association of medical museums—the forerunners of what today are called pathology laboratories—in Boston. There in spring 1936, less than two months after the deaths of the babies with missing right hearts, Helen unveiled her breakthrough.

7   NAKED ON THE BEACH

The preliminary report of her findings to prominent heart specialists in Boston was not the triumphant moment Helen expected. Abbott asked Helen whether it was accurate to say she had found a missing right ventricle in one child since part of the pulmonary artery that arises from it was still present. Rather than missing, wasn't the ventricle simply defective? When Abbott expounded on the differences, Helen gave in to her criticism. "Then it is aplasia," she said.[1]

Abbott was genuinely excited and found Helen's cases interesting. The description of the defective part was a technical point that Helen corrected in the paper she would publish in September.[2] That evening the jovial Abbott invited her for a celebratory dinner. But the exchange with Abbott was dramatic enough to shake Helen's confidence and hurt her feelings. (Decades later, she would still complain about it.)[3] Had it been the only criticism of her work, Helen might have suffered it silently, but her unease with developments in her clinic and anxiety over her future was such that Abbott's seemingly minor critique triggered a full-scale reassessment by Helen of her position and how best to pursue her research. The following week, while at her father's home in Cambridge, Helen considered whether to hand in her resignation. She had discovered that researchers in her clinic were earning more than she was and felt uncertain of Park's support. She sought guidance from the foremost expert, the most prominent woman in science, Florence Sabin, who was now at the Rockefeller Institute in New York City.

Sabin knew firsthand how difficult it was for women academics. She had left Hopkins after being passed over for a job most faculty and students there believed she had earned: chair of anatomy. The famous anatomist Franklin P. Mall, Sabin's mentor, had picked her to succeed him. The job went instead to Sabin's former student. Protests from faculty and students forced the medical school to promote Sabin to a full professor, the first woman. Helen had stayed in touch since medical school. Most recently Helen tried to convince Sabin to head a newly formed Johns Hopkins medical school women's alumnae association, the purpose of which was to shore up the finances of the boardinghouse for female medical students. In the end, Sabin agreed to serve as president for a year if Helen, who had been elected secretary-treasurer, signed the checks.

The letter Helen wrote to Sabin in April 1936 contained a wholly different tone, one of distress. "My situation is this . . . ," she began. In the months leading up to Helen's presentation in Boston, a young female doctor, Frances E. M. Read, had moved into her heart clinic and produced what Helen acknowledged was a "pretty piece of work," revealing a significantly higher incidence of rheumatic fever in families than other diseases, even tuberculosis. The results were so promising that the eminent public health professor who had suggested the research, Wade Frost, renewed the woman's fellowship and hired a second doctor to help analyze data from Helen's clinic and investigate the cause. Now these younger researchers were about to outshine her in an intriguing new field, epidemiology, and she had little part in or control over the ways they used her material.

Their success deepened Helen's frustration over her dual role as patient caregiver and academic researcher. After six years as head of the heart clinic and a heavy patient load, she still had published little aside from a few case studies and a report on how to manage children with rheumatic fever. While Frost's researchers basked in his exuberance and support, Helen endured repeated and troubling rejections from Park when she tried to publish what she considered new and important insights gleaned from her clinic. Adding to the sting was her discovery that these junior researchers, who were paid by the US Public Health Service, were earning 5 percent more than she was. She had broached the salary difference quietly with Park. He recommended her for promotion to associate professor but doubted that it would be approved.

The scrawl of her handwriting and the pace of her sentences revealed Helen's disillusionment as much as her lament over her small list of publications and overwork with patients. "Wouldn't it look to any impartial observer as if I was not measuring up to the J. H. requirements & expectations & if so, hadn't I better leave?" she wrote. Yet Helen believed Park sincerely wanted her to stay, and she underlined for Sabin what he said when she confronted him over the salary difference: "that he hoped that if he gave me an unqualified opportunity for work, I would not ask for more." Her beef was not the salary difference itself but what it represented, that her work was not as valuable as that being done by the others. "My self-respect demands that the institution I work in should have confidence in me," she wrote.[4]

Park said he had confidence in her, but "when I come to publishing anything really new he hesitates & does not want me to," she told Sabin. They were in discussions over what Helen could present at an upcoming pediatrics society meeting in Atlantic City. She wanted to reveal that high blood pressure (hypertension) occurred in children with rheumatic fever. She believed her observations were sound, but Park had pressed her to rule out other possible causes. Helen's article on the management of rheumatic disease was packed with useful information and established her as an expert, but its prescriptive tone was based primarily on her experience treating patients. As considerable as this was, other doctors treating rheumatic fever patients were beginning to advance theories in a more structured way, with controls and hypotheses. Organized clinical research and mass studies were still hard to find in journals, and databases were crude. Hopkins, home to the first school of public health, had set the bar for a scientific approach to public health issues, and its professors were designing mass studies and thinking about how to systematically evaluate data.

Helen's letter to Sabin represented her pique, even fury, about the situation. Yet when she stepped back to consider her situation, Helen knew the risk to her career was high by resigning. Where would she find another boss as supportive as Park? Sabin saw all this clearly. Her response was penned the day she received Helen's letter. Her advice: Go right on with your work and forget about comparing your salary to those of others. She reassured Helen that Park thought highly of her; he had told Sabin so at length. The matter of publication was different; if Helen was so sure of her results, she should try to present her data more clearly to Park. If she could not get his consent, she should get more data and try again. If Park was still hesitant, take full responsibility and discuss the need to get it into print so that others might use it.[5]

Sabin knew Park from his early days at Hopkins and had visited the Parks at their home in New Haven when he ran pediatrics at Yale. He was the rare mentor who could help a woman succeed, as Franklin Mall had given her the project that helped her succeed. Mall, humorously, according to Sabin, had a way of poking fun at self-important men caught up in academe, belittling traditions to reduce their significance for women banned from participating in them. Sabin now shared the tongue-in-cheek advice Mall had given her with Helen.

"Women can get what they want in the way of opportunities to work if they will give up the chance to walk in processions," she quoted Mall in her letter to Helen. Then, in words that she had heard men in this era use to console successful professional women, she told Helen: "The real thing is the work, and the chance to do it and one has to find happiness in that." Eventually, Sabin predicted, the money would catch up. Her advice was professional if somewhat impersonal, although she was said to be a warm person. She did not offer sympathy or refer to her own experience at Hopkins, addressing only the reality, because she knew they had little time to dwell on what they could *not* have or to protest something that would not change in their lifetimes.

Sabin's letter warmed Helen nonetheless. She told Sabin she was grateful for Sabin's sound and friendly advice. In Baltimore after the Easter break, Helen had been greeted by clinical data buttressing her theory about high blood pressure in rheumatic children, and "in my heart of hearts" she wrote to Sabin, she knew she would be a fool to leave. "Your letter clinched the decision." As for her happiness, she said, "There is no question about that. I enjoy my work tremendously & [sic] happy here."[6]

It is the work that counts. Helen repeated these words to herself and would pass them on to other women, although she knew that they were not entirely true. Women, if they put aside their desire to choose their own jobs and win the same affirmation as men—money, title, robes, and fancy colored hats and collars for academic processions and portraits hung when they died—could have a fulfilling life, and this was supposed to be good enough. It meant that women would always be working in the background and under the authority of men. With the academic robe—or acknowledgment as an equal—comes academic freedom—for instance, the freedom to publish without following anyone's advice, even against one's mentor.

Even men like Mall and Park who spotted and advocated for talented women took advantage of their limited opportunities by matching them to jobs that these men needed to enhance their own interests. As Park recruited Helen, the fabled physician Sir William Osler, who later became one of the founding doctors of Johns Hopkins hospital and professor of medicine at its new medical school, prodded Maude Abbott to catalog heart specimens at McGill after she showed him a badly deformed heart. In Baltimore, Osler enticed Marcia Noyes into collecting medical journal articles that doctors could readily consult. She built one of the top medical libraries in the country.[7] Enamored of the work, these women accepted their jobs gladly because, otherwise, life would not be so interesting. The question was not what women wanted to do but what men allowed them to do.

Female professionals, like female factory workers, also could not hope to earn as much as men for the same jobs. Park admitted as much in a note to Sabin in search of a fellowship for a woman forced to work summers to pay her tuition. "The fact that she is a woman, and, therefore, has not got the earning capacity of a man is an additional reason," he said.[8] Women even more than men needed substantial personal wealth to become academic doctors. Helen lived on money from her grandmother and father and, like each of her sisters, left home with a trunk of silver.[9] Abbott, raised by her grandmother and the primary caregiver for her chronically ill sister, struggled financially despite her achievements and her field-defining book.

Helen was at an additional disadvantage as a clinician at Hopkins. Hopkins salaries were notoriously low compared with other medical schools. Only the chairman of a department could become a full professor, Sabin being the exception. There were fewer than several dozen full-time faculty. And the bias toward bench scientists remained despite the conviction by Park and other scholars that to advance medicine, science had to be married to patient care. Finally, Park was constantly strapped for cash to run his experimental clinics; any increase in Helen's salary would have to come from donations.

It is the work that counts. This is what successful women of this era told themselves as they confronted pervasive experiences of injustice and learned to cope with it. As if to acknowledge how wrong this phrase was, Helen updated Sabin on money she helped raise to maintain the boarding-house for women medical students. Among other things, the house needed a new furnace. Helen detailed how she and others recruited one hundred women alumni to help maintain the house—a critical need for women students—via annual dues. "We hope this meets with your approval," she said.

*

Park readily admitted taking advantage of Helen's gender when he set up his clinic. Yet he wanted her to succeed as much as he would any male doctor and endeavored to keep her happy. When she returned to Baltimore after her presentation in Boston, Helen was surprised by a small increase in her salary. It would not change her lifestyle, but it would change her attitude. Park realized this and had convinced Charles Baetjer, chair of the trustees operating the Harriet Lane building, to donate additional funds from his personal charity, the Harriet Lane Community Fund. A lawyer whose family included two doctors on the faculty, one of them a woman, Baetjer was keenly interested in Helen's work and had helped finance her clinic. Grateful, Helen returned to her real mission, advocating for her patients. Her demands were simple: a second social worker, a secretary who could make files in triplicate, and modern blood pressure machines. Unfortunately for Park, she now eyed Baetjer as a potential source of funds.

For months, as she negotiated with Park over her research paper and privately worried over the value of her work to him, Helen had sought an additional social worker. Park refused her request, citing lack of funds. Other pediatric clinics declined to share their social workers. Then one day in early June when Park was away, she ran into Baetjer. Wasting no time, she appealed to him directly. Helen's bold approach upset Park when she told him of it, since the increase in her salary had already stretched the trustees' purse strings.[10] He was more hopeful of obtaining funds from foundations. The annual $5,000 clinic budget paid for salaries, equipment, lab work, and laundry of their white doctors' jackets. He had secured a $1,800 grant from the Milbank Foundation and was in talks for another donation.

*

The heat settled thickly in East Baltimore in early summer 1936, unmoved by ceiling fans that whirled in row house after row house. Inside the hospital, a measles outbreak was finally contained and a quarantine on the ward lifted, but the quiet soon gave way to wretched gasps from the next occupants: children with whooping cough. Helen worried about leaving her patients, particularly a nine-year-old girl at Happy Hills Convalescent Home. She was like a leaf in a storm, battered and blown in and out of the hospital by a cascade of problems, and there was no social worker to visit.

The financial underpinning for Helen's clinic and Park's big gamble on preventive health care became even more precarious. When Park failed to obtain the second grant he had anticipated, Helen realized that another social worker was no longer the problem. Now she feared losing a doctor in her clinic. Angst high, Helen in late June 1936 considered what she called a momentous question: whether to join the US Public Health Service's Army Medical Corps. She did so with Park's support since it might be the only way to save her clinic. The suggestion came from a leading public health advocate and professor of biostatistics, Carroll Palmer. Beginning in 1930, the US Public Health Service expanded its staff of doctors to find the causes of preventable childhood diseases, funding, for example, researchers in Helen's clinic working on the family incidence study. They were part of teams conducting large studies. Palmer liked Helen's study on hypertension and saw her patient records as a source of data. He offered to pay her salary for a year.

With relief, Helen realized she would not pass the physical exam to join the Army Medical Corps and instead took a civil service exam for health professionals that allowed Palmer to pay her from his grant. Her research was now part of the US Public Health Service Office of Child Hygiene's inquiry into childhood hypertension. The family incidence study, too, funded with Frost's government grants, would expand. Palmer even spoke of getting punch cards for the clinic to enter what he hoped would be copious amounts of data. These studies foreshadowed a major increase in funding of academic research by the US government that had begun in World War I to combat soldiers' diseases and chemical warfare.

Helen now had two doctors in her clinic, and one was about to begin testing the blood pressure of more than one hundred children, Blacks in a school and Whites in an orphanage, to establish a baseline blood pressure for preteens. Buoyed, she drafted a bigger operating budget and plotted another overture to Harriet Lane trustees for financial support for her patients when she returned to Baltimore in October 1936. A plainly annoyed Park stopped her, schooling Helen on the limited nature of the charity and forbidding her from approaching trustees without alerting him.

He needed their help for his other clinics. He applauded Helen's ambition for her clinic, nonetheless. "All of this is exactly how it should be," he wrote to her. "Your devoted, time worn friend."[11]

The financial viability of her clinic was so precarious and her future so tenuous that four times in the 1930s, Helen considered leaving for other jobs. She doubted the Public Health Service deal would last; Palmer was disappointed in the paper she was drafting on high blood pressure in children. He needed one thousand children for meaningful statistical analysis. She had identified only forty-nine cases. In May 1937, only seven months after Public Health funds rescued her clinic, Helen applied for a job at Harvard under Park's colleague, Kenneth Blackfan, who ran the pediatric clinic at Children's Hospital. Park recommended her, telling Blackfan he needed women in his department. Blackfan had never hired a woman, and he would not hire Helen because, as he had told Park in 1930 when Park recommended Helen for a residency, the accommodations would fail inspection by a vice squad.[12]

Confirming Helen's fears, Palmer eliminated a part-time position in her clinic in summer 1937. Worried Palmer would pull all government funds, Helen again asked Park to reach out to trustees. But Park had already tapped the trustees' charitable fund to cover a gap in Helen's annual lab budget due in part to her frequent blood tests to detect active rheumatic fever in outwardly healing patients. And after Helen billed the hospital for her new blood pressure machines, Baetjer's charity was under pressure to pay off the hospital's deficit. For now, Park freed his own secretary to duplicate patient files for Helen and promised to beg from Harriet Lane trustees again in the fall. Helen agreed to wait. "I thought that if the situation were hopeless we'd better begin to take in our sails at once but, as it is, we'll gladly let it ride," she told Park.[13]

In fact, that summer and fall, the optimism generated by Roosevelt's New Deal, which had put people back to work through the Civilian Conservation Corps and Works Progress Administration, suddenly burst. Overnight Roosevelt had increased the price of gold, backing the dollar, effectively increasing the money supply. Pumping money into the economy to give people jobs was not working. By mid-1937, businesses lost confidence, and unemployment was headed back up to its high of 1933. In Baltimore, the steel and shipbuilding industries south of the hospital were at a standstill. Women whose husbands had lost their jobs at Sparrows Point steel plant now went to work each morning in canning factories, arriving along with trucks packed with tomatoes, green beans, and spinach from Maryland's Eastern Shore. A discouraged President Roosevelt complained that economists were useless. Helen's father defended his profession, suggesting economists could better predict long-term outcomes. The new Social Security legislation, for instance, was "on the right lines" as far as unemployment reserves and unemployment insurance, Frank wrote to the president, but its old age provisions would "prove difficult, if not impossible to carry out."[14]

Such was the environment Helen found herself in 1937. Without a secure source of funding, her treatments for patients and her research would end. For all her thoughts of leaving, she believed that her work was valuable. Park was exhausted by the search for money and her own, unauthorized advances to wrest funds from trustees. What could she do to help?

Just before Thanksgiving in November 1937, Helen sat down at her desk and wrote a personal check to Park and asked him to use it to make an anonymous contribution to the Harriet Lane Community Fund. The amount of her gift is unknown, but in a note forwarding the donation, Park called it large.

It is the work that counts, even if Helen had to underwrite part of it herself.

There was one other bright spot in the economic relapse of 1937 and 1938: with prices depressed, a local family purchased land to expand its convalescent home for children with rheumatic fever. If this materialized, it would be grand, Helen thought. She wished there was a place like it for "colored" children. Their chances for recovery at home or, in rare cases, in a boardinghouse, were lower than those for children overseen by professional staff. "Some day," Helen told Park, "perhaps we'll have a place for colored children."[15]

*

One reason Helen needed duplicate patient files was so she could take them to Cotuit. Every academic was familiar with the frenetic packing ritual that preceded a vacation. Helen's was planned well in advance. In her car, squeezed between her luggage and pets, in these years a dog named Spot and a cat named Puss, she carefully stacked boxes filled with patient charts and notes. She stayed overnight with pet-friendly academics or family friends and used the opportunity to make and renew connections, attend talks at local medical societies, and detour through the Berkshires. Regularly she stayed at the Park Avenue apartment of her father's Harvard classmate and friend, Charles Culp Burlingham, known as "CCB" to colleagues and "Uncle Charlie" to Helen. The noted labor lawyer, supporter of women's rights, and New York civic reformer was an original public intellectual, serving in these years as a mediator between Mayor Fiorello LaGuardia and corrupt politicians in Tammany Hall. He would introduce Helen to Harvard professor Alice Hamilton and, later, former first lady Eleanor Roosevelt.

In Cotuit, she fell back into the welcoming arms of family. Beginning in 1936, her aging father stayed in Helen's cottage. Catherine arrived from England and Mary from California, bringing their children. With fifteen or twenty family members staying at the Taussig Big House, it was a "mad house," one granddaughter remembered. Helen and her father kept the same schedule, writing all morning and indulging themselves in the afternoon. Frank managed both houses, sending his Cambridge housekeeper ahead to air the mattresses and hiring local men to fix the plumbing and weed the flowerbeds. The staff was reduced to one in the Depression years, and the family pitched in washing dishes. Their food was simple—oysters and clams from the sea and eggs, corn, tomatoes, and greens from neighboring farms.

Helen resumed her early morning laps, usually swimming naked in the waterway sheltered by a spit of land in front of her cottage. This ritual cleared her head of the crushing business of examining patients and writing essay-length assessments in their charts. Space away from the clinic allowed Helen to consider her patients from afar and to confront what for her was the most taxing part of her job: writing. She made so many changes to the first version of a paper that her secretary had to retype it. The editor of the *Bulletin of the Johns Hopkins Hospital*, one of the preeminent vehicles for medical announcements, nearly pulled his hair out over her changes to galley proofs. Helen was so afraid of requesting important last-minute corrections on one 1936 paper that she asked Park's secretary to do it for her. "That paper has hung like a bag of rocks about my neck," she wrote to the secretary.

Helen's hunches about rheumatic fever causing high blood pressure in children were finally taking publishable shape. Unusually tired, writing at the pace of a tortoise, she asked Park for two extra weeks in Cotuit in 1937, without pay, to finish her manuscript.[16]

Harriet Lane was one of the few children's hospitals that regularly took children's blood pressure, and it had done so since 1928. This had allowed Helen to observe that children she had treated for rheumatic fever over the years developed high blood pressure. What caused it? Her investigation uncovered none of the typical triggers. The average blood pressure of healthy children she tested in a school and in an orphanage was much lower than that of her rheumatic patients. Did the infection reveal a child's propensity for high blood pressure, or did a tendency toward high blood pressure make children more susceptible to rheumatic fever? For thirty-seven children in her sample, blood pressure returned to normal three or four months after their disease disappeared. But in twelve other patients, it remained high.

Her paper was the first to show that high blood pressure was an important problem in children. It could be far bigger than anyone suspected, and it might not be related to rheumatic fever. Helen urged doctors to start systematically recording the blood pressure of children; the more they learned about the problem, the more likely they would be able to pinpoint its cause and devise preventive measures.[17]

*

Helen very likely also had in her luggage in the summer of 1937 the files of four patients whose autopsies had revealed a huge hole in the wall separating the atria, the upper chambers of the heart. She had correctly diagnosed two children before their deaths, including a frail little girl with a disturbing boom in her chest. The child had improved on bed rest and digitalis at Happy Hills, but a recurring bronchitis brought her back to the hospital. Helen had long ago discounted rheumatic fever as her main problem and pegged it to a hole between the upper chambers of her heart that allowed blood en route to the body to be diverted into passageways serving the lungs—the pulmonary circulation. The child was short of breath and underweight, and her skin was translucent. While treating her, Helen had remembered a case two years earlier, in 1935, of a boy with a similar enlarged heart, depressed chest, and small frame. She had wavered between attributing his condition to a hole in the heart wall and an open fetal duct or a possible case of rheumatic fever, but in the end, she settled on a defect in the wall separating the right and left heart (atria). An autopsy confirmed her diagnosis.

Discussing these cases with doctors in the pathology lab, Helen learned
of two adults with similarly misshapen hearts. The more she reflected on
the four cases, the more she saw similarities. In life, the adults had been
diagnosed with rheumatic fever disease; the holes in their hearts had been
discovered only after death. The key question was whether the defects in
the adults' heart walls developed as the result of rheumatic fever or were
present at birth. Helen concluded they had been present from birth, since
all the patients showed a chest deformity indicating that the right side of
the heart had become enlarged early in life, when the bones were still soft.
The pulmonary artery was oversized and the aorta undersized, raising the
possibility that the aorta too had been defective from the start.

This was her most sophisticated research finding to date. To publish it,
Helen convinced the doctor who performed the autopsy how a gap in the
wall between right and left hearts changed circulation and affected the
body's development and mechanics. "I've got to present it with such clarity
that Follis [Richard H. Follis Jr., the pathologist] will be convinced of the
truth," she told Park.[18]

The paper she published in early 1938 explained how to diagnose a
patient with a hole in the wall between the two upper chambers—atria—
and how to recognize when an enlarged right heart was caused by rheumatic fever or a hole present at birth.[19] Someone born with a hole in her
heart would be small for her age, not very strong, and pasty. Maude Abbott
had associated the condition with an underdeveloped body in conversation
with doctors who sent her specimens of defective hearts.

Now Helen began to visually assess patients as they walked through
her office door. And in a $5 scrapbook, she began to describe the most
interesting cases of malformed hearts she had uncovered. Her goal was to
publish a book on how to diagnose defective hearts and explain how they
worked. Park liked the idea and offered to find a publisher. Most of the cases
involved blue babies: infants or children whose skin was blue from a lack of
oxygen in their blood. The normal path for blood to reach their lungs was
blocked or partially obstructed. To stay alive, their bodies relied on other
routes, usually other defects, like a still-working fetal duct or a hole in the
heart wall that allowed blood to spill over into the lungs and get oxygen.

The most frequent condition Helen had diagnosed was called tetralogy of Fallot, after the French physician Étienne-Louis Arthur Fallot who
publicized a description of it in 1888. It featured four abnormalities that

reduce the amount of oxygenated blood circulating in the body and alter the balance between the (pulmonary) circulation to the lungs and the (systemic) circulation to the body. The primary problem is a partially blocked or narrow pulmonary valve (pulmonary stenosis) that makes it hard for any blood from the right heart to pass into the pulmonary artery on its way to the lungs. The right heart, because it works harder to try to push blood through, becomes muscular and enlarged (ventricle hypertrophy). The two other defects cause oxygenated and nonoxygenated blood to mix. One is a hole in the wall that normally separates the body's two circulation systems. The opening (ventricular septal defect) allows nonoxygenated blood from the right heart to mix with oxygenated blood in the left heart, reducing the oxygen content of the blood going to the body.

The other is an out-of-place aorta. This master artery is usually attached to the left ventricle to open and accept and move oxygen-rich blood throughout body. In tetralogy of Fallot, the aorta sits above the hole in the heart wall, nearer to the right ventricle, and has a valve that opens to accept blood from both ventricles. As a result, unoxygenated blood from the right heart mixes with blood moving on to the body.

It was miraculous that *any* blood got to the lungs for oxygen. Even more surprising was the way victims of this condition instinctively moved their bodies to try to influence the amount of blood getting to their lungs. Helen noticed that in the crib, babies would bring their knees to their chests when they grew breathless. Older children abruptly squatted. This movement cut off blood to the legs, forcing it to change direction briefly and rush to the lungs to be oxygenized. What she called the "Fallot squat" was so distinctive and yet so unimaginable that from time to time when she was explaining the condition to Park or young resident doctors in these early years, the very ladylike Helen startled them by suddenly dropping to the floor herself to demonstrate it.[20] (Helen would provide the first full description of the phenomenon and show mothers how to move their infants to get oxygen).[21]

*

No matter how often Helen told herself there was something to be gained from studying malformed babies, the air remained pervasive with skepticism. When Helen's friend Kuttner in New York heard about Helen's book, she worried that Helen was wasting her time. Who would read it? Doctors in the hospital, too, dismissed Helen's interest in these forlorn babies. They were happy just to be able to separate a heart defect from rheumatic fever and move on.[22]

Family and ritual sustained Helen in these critical years. Those first summers in her own cottage, she stopped by the Morse barn in the evening for sherry with childhood friends, grandchildren of the diarist James Morse and his recently deceased wife, the abolitionist and paper-cutting artist Lucy Gibbons Morse, whose scenes of children swinging from pine trees Helen so admired that she decorated her lamps with them. Afternoons in her cottage, she practiced lipreading with her niece, Polly, now a young girl, who recounted her adventures, stopping every few minutes so her "very shy" aunt could repeat what she said. In 1937 or 1938, Helen also enjoyed the regular company of another childhood friend, Louis B. Wehle. A labor lawyer who had worked with Helen's father during World War I, he advised President Roosevelt on energy and represented the United States in the Hague. In photographs, he sits on the thin strip of beach in front of Helen's house wearing a white shirt and tie, as if just off the train, smiling and engaged in conversation. A cousin by marriage—his mother was a Brandeis—Wehle was married and made his home in New York, in the same circles as "Uncle Charlie" Burlingham. Polly observed them together so often and so companionably that she concluded Wehle was in love with Helen.

That Wehle was important to Helen is evident; of the many famous friends and colleagues who warranted an obituary in the *New York Times*, only his was preserved in her personal papers. It was dog-eared, along with photos of him. Perhaps Wehle was the kind of man she could see herself marrying. If she had a lover, Helen did not reveal him or her to anyone who betrayed her. In later years when Helen comforted her friend Charlotte over a lost love, she did not draw on her own experience as she was wont to do in other circumstances. The topic was not broached, perhaps because there was nothing to say. About such things, Charlotte would say, one never knows.

But throughout her life, the woman many viewed as shy engaged in intense and varied relationships that deepened her understanding of human nature and fed her mannerly and compassionate responses to other people's feelings and problems. She also actively engaged the natural world, finding meaning and renewal by immersing herself in its physical presence:

swimming in the cold, deep channel in Cotuit beside puffer fish, eel, and minnow; cultivating plants in the dirt; walking the pine woods; and, on the examining table in her Baltimore clinic, sensing rumblings of blood moving inside a patient's body. Every day she went skin-to-skin with a child—her hands on the skin over a chest, liver, or spleen—and regularly she hugged a parent or child. This was a barrier few others could cross in a family's life; it allowed her to diagnose a child but also to establish an intimacy that gave her authority with the patient and simultaneously a personal connection that propelled her desire to help the patient.

For a woman sometimes described as gawky and ungainly, uncomfortable in her own skin, Helen had none of the inhibitions about her body that women of later generations would collectively shed. Her early morning swim in Cotuit provided evidence. Swimming naked dated to her childhood, when her father would take the family by boat to an island off the coast to picnic and swim in the nude. This ancestral tradition long predated the twentieth-century German free body movement that emphasized healthy bodies in direct contact with nature.

Aware that her guests might find her morning ritual unusual, Helen, after she began inviting colleagues and friends to stay with her in Cotuit, devised a way to spare *them* embarrassment by suggesting they not stroll on the beach before 7:00 a.m. As fair warning, she tied a handkerchief to bushes along the path to the water. If it was gently waving, visitors knew to return to the cottage unless they cared to see a tall, thin, white-skinned woman adorned in nothing but fiery blue eyes.

8   GO WHERE YOU ARE WANTED

Helen's confidence in her own methods was the reason for her success but also a source of conflict with others. She had so little regard for the impersonal statistical studies Palmer was developing using data from her clinic that she reviewed patient charts herself for evidence to exclude them. When presented with a paper on rheumatic fever in families prepared under the direction of Drs. Frost and Palmer, professors shaping the field of epidemiology, she rearranged it. From their point of view, she had weakened the manuscript.[1] Resentful, the US Public Health Service team tried to negotiate with Helen. Unable to agree on standards for data collection, Palmer told Park he would drop Helen's family study to focus on the hypertension study or pull out of her clinic altogether. Helen most wanted to continue the family study because it had the potential for immediate impact on patients. But Park advised her to acquiesce to Palmer since they had accepted money from the government and had an obligation to continue. Helen was cornered. The US Public Health Service meant an extra $3,000 for Helen's clinic; without it, Park would have to raise $4,000 for new doctors, although he kindly said he would cut her salary by only $200. Helen accepted Park's recommendation.

This development shook Park as much as it did Helen. Soon after, on June 1, 1938, Helen received a letter from Park, one he instructed his secretary not to copy or keep. Park believed in blunt criticism. He also believed that periodic blowups cleared the air, according to Helen, who took solace from this fact and used it to put the matter in perspective. Their close personal and professional relationship continued. The year had opened with Helen helping to stage a sixtieth birthday party for Park. It would end December 31 with a telegram from Park to Helen in Cambridge advising her that Beethoven's exquisite opera *Fidelio* would be broadcast by the Metropolitan Opera that afternoon. In the intervening months, Park set about finding money so Helen could hire a new doctor and office assistant to continue the family study. He pestered the trustees of Harriet Lane on her behalf. He committed himself to her papers. In letters, Park remained "devotedly yours." For her part, Helen vowed to get over her disdain for statistics and take a class so she could assemble her own data.

Countering this setback was a subtle and exciting shift in Helen's research. Within weeks of the US Public Health Service ultimatum, Helen received Park's comments on three papers she had drafted for publication. Two of them were on malformed hearts—one on how to diagnose transposition of the great vessels using the fluoroscope and the other a description of a child born with an extremely rare thickened left heart wall.[2] She was devoting more and more of her time to malformations as she tried to differentiate the signs of rheumatic fever from a defect that could only have been present from birth. She even began combing through records of patients with hazy diagnoses she believed she could now confirm. Her challenge was to find the right words to explain congenital heart defects so that others could understand and diagnose them. "As I try to put things in clear English, I understand why congenitals are so hard for others. However, I am making some headway and hope to have a good paper eventually," she told Park, who was relaxing at his cottage in Margaree, Nova Scotia. "Good luck, fishing."[3]

*

Pleased with the papers she had submitted for publication, Helen arrived in Cotuit in the summer of 1938 with two tasks: find a replacement doctor to help her finish the family study and pack her bags for a late August voyage on the *Queen Mary* to visit her sister Catherine in England. Helen also immersed herself in finding jobs for German academics temporarily living in other European countries; they were Jews who fled beginning in 1936 when German chancellor Adolf Hitler barred them from professional jobs. Helen's sister Mary, a volunteer in the Boston branch of the New York committee to help displaced foreigners, regularly flooded Helen's mailbox with résumés.[4]

While Helen was away, something happened in her native Boston that would fire her mind for her patients as never before and cement the shift of medical innovation from Europe to America. There, on August 26, 1938, a young assistant surgeon (thirty-three-year-old Robert E. Gross, a long-time correspondent of Helen) opened a little girl's chest to tie off a tiny, superfluous blood vessel connecting the major arteries of her heart. This was the patent ductus arteriosus, the fetal passage that, when it failed to close after birth, allowed blood intended for the body to leak into the lungs. It was less than a quarter of an inch long and three-eighths of an inch around and was disposed of with a single no. 8 braided silk suture.

News of the successful surgery on seven-year-old Lorraine Sweeney spread quietly at first, because the *Journal of the American Medical Association* (*JAMA*) refused to publish articles that had been leaked to the popular press. But the five months between the operation and the publication of Gross's report in February 1939 was very short by *JAMA* standards.

When she learned of it, Helen was momentarily stunned. Rifling through her brain's catalog of patient files and illustrations, the ones she was collecting in her scrapbook and what she had learned over the years from the jars of hearts in the laboratory, she realized immediately where his operation could lead. She was thrilled that patients in her own clinic with an abnormal duct like Lorraine Sweeney's might now lead normal life spans with Gross's surgery. But they were not the ones who came to mind when she read Gross's report. Helen had diagnosed far more complex and fatal heart defects, including in babies whose entire circulatory systems were reversed: aortas had developed where pulmonary arteries to the lungs should be and vice versa. She had deduced how these defects altered the circulation systems and described the Plan B operating system that kept patients alive. The body's ability to develop a Plan B in infancy meant that in an emergency, existing parts, such as the patent ductus arteriosus, played unusual roles.

As a result of her research, Helen was one of few people who knew that the patent ductus arteriosus that Gross stitched shut in Lorraine Sweeney because it let too much blood into her lungs was in other children the only thing keeping them alive. Many blue babies—those with reduced oxygen because the normal path to their lungs was blocked—depended on alternative routes like the open fetal duct to get *any* blood to the lungs to pick up oxygen.

She even explained it to the chief of pathology. At a conference in his laboratory one day in 1939, William MacCallum grew flustered during an autopsy as he tried to trace how the child had lived, given his circulation system, in which the passageway to the lungs was blocked. Helen suddenly spoke up. She drew his attention to strands of fetal duct tissue, indicating the duct had been recently working. Once the fetal duct closed, eliminating the only route for blood to the lungs, the child's circulation was no longer compatible with life. MacCallum, who ignored women doctors, appeared astonished. Neither he nor others in the room responded.[5]

Helen had wondered when she discovered the importance of the open duct for children with reduced oxygen if there might be a way to keep it open permanently. Could a chemical reaction developed in the laboratory keep it open? But she had not the time to investigate.[6] With Gross's operation, she thought of a different way to save these children: build a new duct to bring blood to the lungs, where it could be oxygenized.

If a surgeon could cut open the chest, reach in, and stitch closed an errant blood vessel diverting blood to the lungs, couldn't a surgeon build one of these vessels to bring more blood to the lungs? Could blood vessels be moved around? She knew little about surgery. But the flaws in her patients with malformed hearts seemed to her like a plumbing problem. She would need to get in touch with her colleague to ask him if he could help her save children with more complicated problems than an open duct.

*

By summer 1939, Helen's association with the US Public Health Service and her government salary ended abruptly when Palmer moved his grant money to a project in New York. Park promised that, from now on, he would arrange for the Harriet Lane trustees' charity to pay her entire salary, but Helen, not Park, would now plead for outside funds to pay the salaries of researchers in her clinic. The outlook was poor.

Grants from major foundations then financing the bulk of academic research in the United States dropped dramatically during the Great Depression and would become smaller still as the United States made war preparations. Any money to be had in the late 1930s went to a suddenly fashionable hot spot: psychiatry projects involving children. In the toughest financial climate ever for Helen's clinic, the Rockefeller Foundation poured money into studies whose outcomes could improve public health.

Hopkins's public health school itself won the huge sum of $350,000 in late 1939 to establish a department of preventive medicine. As Helen's father had predicted, the field of public health was booming.

She appealed to the Commonwealth Fund with little hope, knowing the fund already supported long-term studies on the cause of rheumatic fever by Wilson in New York and provided most of Jones's $40,000 annual budget in Boston. Jones was engaged in a large study to identify criteria to diagnose the disease.

Helen had advanced rheumatic fever treatment. Once she discovered that regular doses of aspirin reduced inflammation, she tested seemingly well children multiple times after the drug was withdrawn to try to detect a still-active infection that would keep a child in bed. Her careful observations in the early 1930s saved some children and extended the lives of others. She had in 1939 finally published her findings.[7] She also linked some symptoms other doctors thought indicated rheumatic fever to hearts damaged from birth. But her studies were limited compared with the goals of well-established researchers and her time divided.

The sideline research she developed with children with malformed hearts was as unappealing to foundations as it was to Helen's colleagues because it had no practical value. Park believed her descriptions of malformed hearts and how to diagnose them were important enough to publish, but even he admitted in a 1938 grant application for Helen's book that few would read it.[8]

Helen continued to solidify her diagnoses by reexamining current and former patients. Some original diagnoses held, but from 1939 throughout the 1940s, Helen uncovered congenital defects in children previously diagnosed with rheumatic fever. One afternoon in her clinic, Helen reexamined a boy in whom she had previously diagnosed an open fetal duct to be sure she had not mistaken the sound for a hole in the heart wall. Except for whooping cough and a slightly enlarged heart, the boy had remained well for five years. Someone with his condition could expect to deteriorate in early adulthood. For the first time, Helen could suggest a procedure to give him the chance of a normal life span. He was a candidate for Gross's operation.

*

After mailing her pleas to foundations, Helen departed earlier than usual for Cotuit. Worried that she felt hamstrung by financial troubles, Park wrote to reassure her and advise her not to work over the summer. "Was there anyone ever as dear & as kind as you to write me right after I'd run away?" she replied. She reassured Park in return: whatever happened, the improvements to her office space in the clinic would be enough to cheer her for the fall. She regretted he had to take on any work at all regarding the grants they were seeking for her studies. As for his advice not to work, she rejected it: "I should grow restless with a huge vacation and feel dissatisfied should I return with nothing accomplished."[9]

Patient files took over Helen's upstairs bedroom, awaiting her analysis for the paper she was writing with Marcel Goldenberg, a Jewish academic at the University of Vienna who had fled to London to work with Lewis. Charts of children's weights and other measures covered the bed and every available space on tables and chairs. The floor was obscured except for a path to her sleeping porch. Helen had laid down X-rays of infant hearts side by side for visual comparison. This was the scene that Helen's distant cousin, Marjorie Prichard, encountered when she arrived at Helen's cottage in August 1939. As Helen led Marjorie into the room to discuss her research, her dog, Spot, and Puss, her cat, followed them and sat down to observe. Prichard was charmed.

Her visit came at a crucial moment for both. Each woman treaded carefully in environments controlled by men. A physiologist at the University of Oxford, Prichard was an expert on the circulation of the fetal heart, including the ductus arteriosus; she was trying to pinpoint when after birth it closed, a subject of great interest to Helen. Part of her research involved dissecting fetal hearts and making films of animal hearts.

Helen tapped her expertise to help read the X-rays on the bedroom floor. It is easy to imagine the pair discussing Helen's idea to help blue babies following Gross's surgery on the fetal duct and the scrapbook she hoped to turn into a book. It would not have been surprising if the women shared their anxieties about the future, Helen's because of the loss of funding and Marjorie's because of ominous events in Europe. England was poised for war, having pledged to defend Poland if Hitler continued to march his troops across the continent.

In the beautiful setting of Helen's cottage, the women mixed business with pleasure. Armed with flashlights, they wandered the beach until midnight, observing the lives of sand hoppers, the tiny fleas that when roused would fly up into the air to ward off danger. The water gently lapped the sand and their own bare feet.

After leaving Helen, Marjorie spent three hours in Boston with the radiologist Merrill Sosman, an expert at diagnosing diseases with X-rays at Harvard's Peter Bent Brigham Hospital (now Brigham and Women's). Sosman was so impressed by her discussion of the circulation of the fetus that he asked Marjorie to lecture his assistants. "I really enjoyed it," she wrote Helen, ". . . in spite of my consciousness of the paradoxical situation of little me expounding to the medical profession."[10]

While Marjorie was in Boston, on September 1, 1939, Germany invaded Poland. In response, Britain and France declared war on Germany. Unable to see Helen again in Cotuit, Marjorie deposited her 16 mm films on circulation systems with Sosman for safekeeping before she returned home.

Marjorie's report of her warm reception at Harvard by itself would have been enough to set off Helen's competitive gene. She did not yet know that one of the doctors so keenly interested in Marjorie's lecture on fetal circulation was the surgeon Robert Gross.

Seven months had passed since Gross had published an account of his operation. Helen had hung on to her idea for a surgical solution to bring more oxygen to children with malformed hearts. She did not seek others' opinions or discuss it with Park, as far as we know. Sometime in fall 1939, emboldened by Marjorie's success and impatient with the lack of her own, Helen made a special visit to Boston Children's Hospital to try to sell her idea for saving blue babies to Robert Gross.

*

Helen Taussig and Robert Gross were kindred souls. Like Helen, Gross was shy. He too had overcome a physical limitation that forced him to develop superior technical abilities and good judgment. He was the rare surgeon who had first completed a residency in pathology.

Her diagnostic skill was so well established that few people knew she had lost much of her hearing. Gross's surgical skill, too, was so widely regarded that few people knew that he had sight in only one eye. He had a congenital cataract, and from childhood, worked to increase his depth perception by repeatedly taking apart and putting together watches.

To surgeons, the heart was off-limits. Few with a knife dared approach, much less enter, the heart, the electrical pump that circulates blood to the body. They confined themselves to forays into the stomach and gallbladder, the extremities, neck glands, and, in cases of tumor or trauma, the brain, including on soldiers in World War I. From the 1800s forward, experiment after experiment on the human heart failed. In the 1920s, two doctors, including Elliott Cutler at Peter Bent Brigham Hospital, had attempted to operate on adults to repair a mitral valve, the valve that opens and closes to let blood flow between the upper and lower left chambers of the heart. Six of seven patients died. These failures convinced doctors that Helen's effort to diagnose heart defects was useless.

Robert Gross interned under Cutler. He had read Cutler's prediction that the first successful heart surgery would be to tie off the problematic fetal duct.[11] As a resident in pathology in the early 1930s, he performed autopsies of children who died because of this appendage. By 1938, after tours of European hospitals, he was at Boston Children's Hospital as chief surgical resident (under William Ladd, the first chief of pediatric surgery at any hospital). Gross and a partner experimented with ways to close off the fetal duct, first in the pathology laboratory and then on dogs in the surgical research laboratory. They decided on a simple suture, or stitch. Then Gross would knot it, like he was tying a shoelace.

Compared with Cutler's failed attempts deep inside the chambers of the beating heart, Gross's operation was easy. He was cutting off a blood vessel outside the heart and stitching closed the spots where it joined two larger vessels. Still, there were unknowns that could not be predicted in the laboratory. The duct linked the aorta to the lungs, diverting blood to the lungs. What if blood didn't easily reroute itself through the patched aorta and out into the body once the duct was removed? Would the patient, little Loraine Sweeney, survive anesthesia? Gross was certain his operation would help this child with troubled breathing. He was also certain that his boss would not allow him to try it, so he waited until Ladd was out of town and obtained approval from a deputy.

Lorraine recovered within days. Gross's removal of a misplaced pipe restored normal blood flow throughout Lorraine's body. Instead of certain death at an early age, she would live into her eighties.

Gross's future was secure. Helen's was precarious. As she walked up the steps into Boston Children's Hospital and rounded the hallway in search of Gross's office, Helen knew the stakes of this visit were high.

Helen had often revisited the city of her birth to make connections and exchange scientific knowledge, ever mindful of twin goals: helping her patients and advancing her career. Her clinic's financial problems, her research on congenitally defective hearts, and, above all, her desire to help children with seemingly intractable problems had brought her here once again. If she could convince Gross to work with her, she might finally return to her beloved Boston and her father. She also might move forward on behalf of her patients.

Because of her focus on congenital malformations, Helen had begun to attract the sickest children, the extreme cases, the most desperate families. Once diagnosed, these children's deaths were preordained. Helen knew why and how death would come. Often she was with her patients when they died and afterward consoled their parents. Then she examined the child's heart in her hands. This made her an expert in tragedy and one energized by sorrow. The problem was no longer the defective heart but what to do about it.

Removing an extra vessel from an otherwise normal heart, as Gross had done, was different from installing a new vessel in the same spot. But Helen considered Gross's surgery against what she knew children with defective hearts needed, not what she knew about surgery. She envisioned creating another abnormal structure that could work with a flawed heart to create a new normal.

Surgeons had been able to stitch together slashed blood vessels and veins in the stomach or seal the skin over amputated limbs since well before 1902, when a French surgeon, Alexis Carrel, son of a textile manufacturer, perfected the method with tiny needles and waxed embroidery thread. Using such needles and Carrel's techniques, surgeons had also sewn tissue from one part of the body into another. But could they build a blood vessel in a new place? How? Could they repurpose an existing vessel? Which one? And how would the body react to this attempt to move around pipes that carried blood and nourished it?

Helen did not know exactly. Despite years of experience observing autopsies, she had spent little time inside an operating room. No one knew much about the type of surgery she was suggesting. To Helen, the "how" was a problem to get around in the laboratory, where some had already experimented. She believed a skilled surgeon could figure it out.

In Gross's office in autumn 1939, Helen would have greeted him warmly and congratulated him as they discussed his operation and problems of children with flawed circuits. Perhaps Helen reminded the Baltimore-born surgeon of their conversations beginning eight years earlier about the fetal duct. She was discovering the usefulness of this circulation defect in some patients, and he was a pathology resident cutting into children who died because of it.

The formalities covered, Helen summoned up courage to broach the reason for her visit. Had he ever *created* blood vessels in the laboratory? Was it possible?

According to Helen, Gross claimed to have built "lots of ductuses" or passageways for blood, in the animal laboratory. He did not go into detail, but after Lorraine's operation, he had repurposed blood vessels in dogs to create the very artificial fetal duct Helen sought; they helped him understand how blood flowed in children with the extra passageway that he was closing off.

Then, "rather timorously," Helen suggested to Gross that these little ducts might be a big help to blue babies—cyanotic or oxygen-deprived infants and children who could hardly breathe.[12] She wanted him to establish a direct link between the aorta and the pulmonary artery so blood could move to the lungs: a replica of the open fetal duct. Gross seemed amused by her suggestion. Flatly, without consideration, he dismissed her idea in a way that Helen took to mean it was foolish. "Madam," he replied, referring to his surgery on humans, "I close ductuses, I do not make new ductuses."

Later, Gross would wish he had listened to her more carefully, and he would offer their conversation as a lesson for young surgeons.[13] Gross was so fully absorbed in the success of his work closing off tiny blood vessels that he could not see the opportunity before him to use the test vessels he made in his lab for another purpose. He was a fixer. Helen was a visionary. Both were competitive, but her focus was on finding a solution for her patients.

That evening, Helen returned to her family home in Cambridge to dine with her father. Perhaps over the habitual glass of sherry before dinner, she recounted the details of the day, her proposal to Gross and his swift dismissal. Her father's response was a variation of advice he had often dispensed as Helen sought success in an unwelcoming environment: it is better to stay where you are wanted, he said, than to go where you are merely tolerated.

This time he was right. She knew when she departed Gross's office that her attempts to find work in Boston were over. However chagrined and disappointed she might have been by Gross's reaction, Helen now knew her idea was possible. Her next task was to find a collaborator who would work with her to save a patient with a far more complicated problem than the one Gross so skillfully surmounted. This person would need to examine the contents of the jars she had amassed in rows on shelves in the pathology department: the hearts of her patients. He would need to experiment in a laboratory. All she could do now was wait or, as Helen put it, "bide my time." To cheer herself, Helen drove out to Cotuit to test a newly installed heating system in her cottage. Now she could visit Cotuit even in winter.

Two things occupied her back in Baltimore: keeping her clinic going in a difficult atmosphere and her book. During 1939, the three doctors in the children's heart clinic handled 2,100 visits for 1,945 patients and gathered material for their research projects. In any given month, sixty to seventy children were on bed rest. Most had rheumatic fever. To care for them, Helen worked the system. If sometimes she forgot to seek Park's permission to bill the hospital for extra expenses, such as more blood tests for rheumatic patients, she sent an apologetic note, offering to take the blame should anyone question it.

At the end of 1939, after Park appealed again to the Commonwealth Fund, Helen learned she had won a small grant to start her book. At Christmas she shared the news with Marjorie, one of her few cheerleaders, who congratulated her. Helen also wrote of her futile visit to Gross. Marjorie enjoyed hearing about it, she told Helen, particularly since the surgeon had attended her own talk at Harvard. This had to be comforting.

A few other surgeons had begun to close off the bothersome fetal duct in children in the wake of Gross's success, but Helen was unaware of them. No one in Baltimore approached his interest in the heart; they focused on the brain, pituitary glands, and nerves. A search had begun for a new chief of surgery following the incumbent's resignation after a stroke.

Helen was so sure that the Hopkins surgical staff could not handle chest operations competently that she fended off pressure from interns and the assistant resident to begin Gross's operation. They correctly suspected Helen had candidates for the new surgery, but both Helen and Park kept quiet.[14] There was no rush: children with an open duct lived years with it.

She kept her head down during 1940, consumed by patients and her book, until one day in November, she learned by phone that her father had suffered a heart attack. Helen rushed to Boston to direct his care. For a week he lay unconscious at Mary's house, where he died, Helen at his bedside.

Frank Taussig was the most important person in Helen's life, the one who had regarded her as his intellectual combatant and who championed her grit, compassion, and determination. The first to comfort those in mourning, Helen assuaged her own grief. Her note to Marjorie containing the news was waylaid months, mail being interrupted by sinking ships and the nightly bombing of London. Marjorie understood the blow Helen had taken. "I *do* feel for you so deeply in what I know must be a great sense of loneliness," she wrote to Helen.[15]

A month after her father's death, as Helen prepared for her first Christmas in Boston without him, she learned that Hopkins offered the top surgeon's job to Alfred Blalock of Vanderbilt University. He was a compromise candidate. His most important work was in the laboratory, on the nature of shock, and some Hopkins surgeons were skeptical of his surgical talent. But he was one of the few surgeons who had successfully closed a patent ductus arteriosus after Gross's success. He also had operated on the thymus, a gland underneath the breastbone critical to the immune system. Helen had not read his recent paper on diseases of the heart and possibilities for surgery, but she heard enough to fill her with hope: unlike his predecessors, Blalock was interested in surgery in the chest.

If this man takes the job, she thought, he would be "my chance to get ahead."[16]

9 ON THE PRECIPICE

Slowly the center of gravity was shifting to Baltimore. Some of Helen's patients, regulars in her clinic for more than a decade, were like family. For the first time, she had the prospect of working with a surgeon interested in the heart. Her duplex on East Lake Avenue, while adequate for visiting academics and lovingly maintained, hardly compared to her father's house, but she had an inheritance now; she could afford a house of her own. For the time being, Helen arranged with her siblings to take delivery of a key symbol of their life in Cambridge: her father's dining room table. Her guests would now sit at the table where her father engaged some of the most creative minds of the Progressive era. Helen claimed another heirloom, one her father had used to entertain visitors to his Harvard office: an exquisite sterling silver tea service.

Park, when he congratulated Helen on a major finding or published paper, habitually told her, "You are your father's daughter." She had the same discerning nature and thirst for new knowledge that had allowed her father to carve out a unique field—economics—and achieve a singular reputation. She emulated his work ethic, his love for nature and children, and the way he masked toughness with sweet grace. But it had taken her far longer than her father to arrive at the precipice of success. With neither the opportunity nor the forum her father had enjoyed very early in his career, hampered from the start by her gender and dyslexia, hard of hearing, and at any moment's notice called on to review the care of hundreds of children in her clinic, achieving anything near his success would take more than work ethic or gifts of the mind. It would take willpower far exceeding her father's. She had it.

Helen was creating her own professional space, guided by her own quirky personality and sense of duty. To others, she maintained that her life's work was to treat children with rheumatic fever and, above all, to prevent the disease from progressing to cause death or lifelong disability. Her studies on that disease continued. But she knew she was not among the top researchers trying to isolate its cause. She had veered off into a risky pursuit, diagnosing heart malformations, because she sensed an opportunity that no one else deemed desirable or fruitful. Her book would show for the first time how to diagnose the problems of these tiny hearts from actual symptoms like shortness of breath and subjective signs like heart murmurs and the contour of a child's heart.

The academic pursuit thrilled her, but it was not her primary driver. Her fervent goal now was a surgical fix for the most common heart defect in her patients, tetralogy of Fallot, a problem defined by a blocked path to the lungs. In this, Helen's stubbornness was a boon. Even if she could corner Blalock in his early months at Hopkins to explain her idea or get his attention long enough to present a patient who could benefit from him building a duct like the one Gross had closed, it would not happen soon. Coincident with Blalock's arrival at Hopkins, a different problem commanded the focus and energy of all who might have paid attention to her puzzles: a looming war and a race to improve battlefield medicine.

*

While Europe reeled from Hitler's bloody invasions, Helen had a firsthand account as the horror played out in England. Composed during a lull in air raids, Marjorie's sympathy letter to Helen in January 1941 recounted diabolical events, mass migration, and interrupted lives. But Marjorie also described her radiology work with dogs and her ongoing discoveries about fetal circulation. Her list of animal hearts and spleens she had dissected included an elephant from Africa.[1] Meanwhile, Helen had patients to fret over. With no empty hospital beds one night, she commandeered a cot and stuck it in a utility room to accommodate a very ill child whose impoverished family had traveled from West Virginia for help. She examined the girl and met the family for the first time in the converted closet.[2]

Pressure on Helen came not only from desperate patients; she also confronted a challenge to the patient-centered culture Park instilled in his clinics. For over a year, prior to the surrender of France in spring 1940, Hopkins had quietly begun preparing two hundred doctors for battlefield medicine at the request of President Roosevelt. Helen lost staff and competed for supplies as colleagues commandeered labs and lecture halls to graduate doctors in three years instead of the usual four.

The atmosphere outside the hospital was charged as well. Workers by the thousands arrived for jobs at Sparrows Point, where Bethlehem Steel had opened new slipways or steel docks—doubling and tripling the number of "Liberty" ships they could build and lend to England to replace those scuttled by German submarines. Baltimore was among the cities erecting unoccupied house-like structures along the shoreline, dummies to divert the enemy and reduce civilian casualties. Fear was palpable. On balmy nights, crowds gathered in a city plaza to hear speakers denounce Hitler.

In the face of death on the European battlefield and in her own clinic, Helen could do nothing but practice patience. She also demonstrated more than her usual compassion.

She worked Sundays now, so it would be understandable if her correspondence with Marjorie suffered. But Helen never abandoned her cousin. Instead, she became a more faithful writer. In their regularity and content, Helen's monthly letters displayed an empathy she typically offered those in crisis. Helen's cheerful recounting of the cousins' past adventures in Cotuit provided the grateful Marjorie with some "fun, a pleasant touch in the all too grim, & serious atmosphere."[3] Besides moral support, Helen tucked a botanical print into one of her letters. Another time it was a packet of New York no. 12 lettuce seeds for her cousin's Victory Garden. Marjorie had requested the supersized variety after a friend had had "magnificent" luck with it.

The benefit to Helen from this relationship was normalcy of a different kind. She had Park solidly behind her. But in a sea of doubt, the presence of another woman scientist interested in her work provided affirmation she did not receive from peers. Marjorie was roughly the same age, of similar background with similar interests, who also hoped to showcase her original research in a book. The bulk of the cousins' correspondence centered on their scientific research and discoveries. Both women had attacked problems of physiology, that is, how the heart and blood circulation worked. Marjorie was examining the earliest stages of life, notably the development of the duct that aided fetal circulation; Helen, how improperly developed or damaged parts—especially the duct—developed or adapted after birth. They shared their discoveries and congratulated each other on published works, but this thread between them also provided inspiration, a chance to learn from another woman working in a high-powered, male-oriented arena, and even a little competition.

Both faced extreme pressure to get their work done and manage their personal lives. We do not know if Helen's house manager took on a wartime job like many women in domestic service did, but with only part-time help in her Oxford cottage, Marjorie hosted three nurses who had fled London, maintained a vegetable garden, and forfeited lab space to the British Army, which had taken over her hospital. Marjorie's accomplishments under the circumstances seemed remarkable. By summer 1941, she was quietly rewriting a book on fetal circulation to be published by her bosses, Drs. A. E. Barclay and Kenneth L. Franklin. Excitedly she told Helen she would be listed as the third author. Information obtained from Marjorie's films would become part of a seminal book on fetal circulation.[4]

Helen managed a few weeks away in Cotuit to write. She also gathered the most compelling patients in hopes of discussing them with Alfred Blalock, who was settling in at Hopkins. To convince the new surgeon to change blood flow to save her blue babies, Helen would need a strategy, and she would need to execute it flawlessly. Beginning in 1942, when Blalock undertook the simplest operations on children, she began to educate him in clinic visits and patient conferences arranged by Park about the problems of blue babies. Her campaign had begun.

In professional but not-so-subtle ways, she pressed Blalock on her goal. She introduced the chief surgeon to gasping patients he could save. She showed him how they squatted to push blood through abnormal paths to the lungs so they could breathe. At one point she offered him an idea—to build a vessel that would bring blood to the lungs—that became the solution. All the while, she offered a promise to her patients that help was in the offing.

*

In her description of the beginning of modern heart surgery, Helen would say that Gross unlocked the door and she pushed it open. Actually after Gross shut her down, Helen pushed right past him. But first, as befitting a proper Boston lady, she knocked.

When and how Helen Taussig met Alfred Blalock is not recorded; Blalock was hardly visible in those early transition months. His focus was not chest surgery. He assumed his chief of surgery post in late June 1941 with what amounted to a second job: chair of a key US Army committee to examine the cause of and treatment for shock to the body that followed a traumatic wound inflicted by bullet or bomb. Throughout the 1930s, Blalock had published path-breaking studies showing that the shock to the body from such an injury, rather than the injury itself, was what proved lethal. Shock was not caused by a chemical reaction as Harvard physiologist Walter Cannon had theorized, but rather a flight of blood away from major organs— heart and brain—to the site of the wound. (It was not loss of blood itself.) The solution, first demonstrated during World War I by George W. Crile, who later founded the Cleveland Clinic, was to resupply vital areas of the body with blood before they stopped working. (Crile performed the first blood transfusion between humans.) Still unknown was how much and what component of blood—whole, plasma, or packed red blood cells— should be pumped into critically ill soldiers. Blalock was pressing the Army to allow him to test his theory on mass casualties in Europe. Instead, the Army asked him to conduct laboratory experiments related to the deaths of civilians in nightly bombings of London.

While Helen attended to patients amid war-related shortages, Blalock fumed over delays. His laboratory fell shockingly short of what he and his laboratory assistant had developed in Nashville to recreate medical conditions in dogs and try procedures on them. Modernizing would take months even if they could acquire steel and other materials being diverted for pre-war preparations.

If the timing for a collaboration was off, Helen at least had an introduction. Blalock had few friends in his early days. His slightly offhanded manner made him seem immature to storied physicians and their elite clientele, and his success had come mainly in the laboratory. But the brain surgeon Walter E. Dandy, then one of the most celebrated surgeons in the country, became Blalock's advocate and tennis partner. Blalock had helped care for his patients in summers during medical school, and Dandy was an old friend of Park; they had interned together in 1914, Dandy in surgery and Park in pediatrics. Before Blalock even arrived, Park had written to him, lobbying him to begin operations on children like those being done in Boston on tumors, obstructions, and open fetal ducts.[5] Despite opposite styles— Blalock wanted to dominate, and Park saw himself as a servant, the quiet agent of others' success—their respect for each other was evident. After

Dandy's propensity to rush into the operating room ill prepared, Park was relieved by Blalock's careful approach to surgical problems. Blalock would say that Park was one of the finest gentlemen he had ever met.[6] So when Park invited Blalock to the weekly conferences he held in the auditorium or on the hospital wards to discuss interesting cases, the very busy Blalock showed up. This connection and the cold reception from other doctors provided an opening for Helen, and she took advantage of it.

*

Helen and Alfred Blalock had briefly crossed paths years earlier. Helen had arrived for medical school in 1924, and Blalock left Hopkins shortly after for Vanderbilt, having served as an intern and assistant resident in general surgery after being denied a surgical residency under William S. Halsted. (Blalock excelled in surgery, but others had higher grades.) A native of Atlanta, he married a Nashville socialite, Mary O'Bryan, and they had young children. His southern roots loomed as large in his consciousness and persona as Helen's New England roots did in her.

But while Helen's father heaped praise on her, Blalock's father was strict and sparing with affection. Blalock attended a military academy. He entered medical school in part because of his father's high regard for Hopkins after his own successful treatment there. In medical school, Blalock worked hard and partied hard. He was known as a ladies' man and as a leader who could build a team, corral troops, and win an athletic contest. Like Helen on a tennis court, he exploited all the angles. In Nashville, Blalock had been disappointed to be put in charge of an animal laboratory, but he soon saw its advantage. His diligence was pronounced, but unlike Helen, who dirtied her own hands to make discoveries, he found and deployed talented others to conduct experiments that would bring him fame.

Central to Blalock's success was Vivien Thomas, the Black man he hired in Nashville to conduct gruesome experiments on dogs to prove his theory on the nature of shock. (Since "Thomas" is not a name by which he was recognized—he was addressed as Vivien, without the honorific afforded White people in lesser jobs or learning from him—I identify him here by both names.) The son of a skilled carpenter and construction company owner who worked his way into the trade, Vivien Thomas's hope to enter college had been dashed by the 1929 stock market crash and the loss of his job building houses. Periodically during his years in Blalock's employ, despite negotiating salary increases, he worked second jobs including as bartender for doctors' parties to pay his mortgage and feed his wife and two daughters.

That a southerner distantly related to the Confederate president Jefferson Davis would hire a Black laboratory assistant, bring him to Baltimore as a condition of Blalock's appointment, and house him temporarily in his own home during the move was revealing. Blalock would have been hard-pressed to find a White man to take a full-time job killing dogs or one that would have approached Vivien Thomas's skill with tools or his ability to analyze data. Vivien Thomas got the job operating on dogs for the same reason Helen got the job caring for dying babies: a White man would not have stayed in it long. Most doctors training in surgery left the lab after a year. It was at Vanderbilt that Blalock organized a system and developed a style that in Baltimore allowed him to build the surgery department into a powerhouse that for several decades trained the country's top surgeons.

Vivien Thomas had his conditions. Two months into the job, after making an error, he found himself being dressed down by Blalock in an obscenity-filled temper tantrum. When he learned such rages were common, Vivien Thomas took off his lab coat and went to Blalock's office to ask for his outstanding wages. He told Blalock he "had not been brought up to take or use the kind of language he had used." Blalock promised never again to speak to him that way, a promise he kept for thirty-four years.[7] This was an exceptional response to someone who challenged his authority, but Blalock needed Vivien Thomas. Besides conducting Blalock's experiments, he taught Blalock the operation that made him famous. He also taught surgery to Blalock's residents.

By late fall 1941, Blalock's revamped laboratory was open for business, and most of it was Army driven. Vivien Thomas began crushing the limbs of dogs to mimic the civilian experience in bombed-out London, then pumping them with blood to try to save them. Such work could save American lives if Germany began bombing East Coast cities.

Meanwhile, to Helen's frustration, her patients' tiny hearts continued to end up in jars in the pathology laboratory. She had been away when a baby boy born at the hospital immediately turned an intense blue. An autopsy showed the baby had complete transposition of the two great arteries, the pulmonary and the aorta. Babies with problem hearts usually did not develop signs of distress for several months, when their fetal ducts finally closed off, eliminating their main source of oxygen. Why had this child's fetal duct closed so soon?

This was the kind of problem she was trying to make sense of when the Japanese navy attacked Pearl Harbor on the morning of December 7, 1941. The next day, after Roosevelt delivered his "day of infamy" speech, Congress declared war on Japan and Germany. By April 1942, when train cars of doctors left Baltimore for duty overseas, much of the hospital was in the hands of Helen and other women, some of whom returned to work for the first time since their marriages.

*

Blalock too was at a crossroad. Relying on slowly transmitted written reports, he found it hard to assess how much blood, and whether plasma or whole blood, to give injured soldiers. With so many staff at the battlefield, Blalock had had to reassign a doctor working in the animal lab to operate on patients, leaving Vivien Thomas alone. The work he had begun in Nashville on the thymus gland and its role in myasthenia gravis disease was waning; he had stopped operating when it became apparent that the operation helped some people but not others. (The condition mimicked the effect of nerve gas used on World War II soldiers. It blocked transmission of the nerve to the muscle, immobilizing it.) He needed something novel to secure his reputation.

At Park's suggestion he would soon tackle what many surgeons believed would be the next heart surgery to succeed: an obstruction in the aorta that caused tragic development problems in children. Like the fetal duct, the aorta was outside the heart, but it was the main blood vessel to the body. Repairing it would not be as easy as Blalock hoped. The risk of pioneering a new surgery had been vividly illustrated by the death of Gross's twelfth patient. The teenage girl dropped dead suddenly while dancing in the living room at a party to celebrate his operation to close her fetal duct.[8] An autopsy showed that the tie on the duct had loosened, creating an opening, effectively a gash in the aorta, that allowed blood circulating to the body to leak in a massive hemorrhage. Gross stopped knotting off the duct like one might tie a shoelace and advised his peers to stitch both ends of the errant vessel tightly closed so that no blood could escape.

Helen had a backlog of patients who could benefit from Gross's operation. These were patients she and Park had shielded from inexperienced residents. At a conference in Helen's clinic in mid-1942, she presented several of them to Blalock, who immediately agreed to begin operations to close the fetal duct at Hopkins. He had already performed three such operations in Nashville and had the benefit of Gross's adjustment as well as ideas of his own about how to permanently disable the extra duct.

Months passed while Helen agonized over the choice of patient. There were no studies yet on the quality of life after this operation or the survival rate. The candidates were not on the brink of death. They could live a good life, if a shorter one. How could Helen justify asking a parent to take such a risk? Finally, she and Blalock chose a "mentally retarded" child for the first operation, and his parents agreed. The selection of a patient with impaired mental ability for experimental surgery rarely raised ethical questions. At this time, human subjects including the disabled and imprisoned were often used for experiments, sometimes without their consent.

On the day of the operation in November 1942, Helen had been a heart doctor for more than twelve years. She had examined with her hands more than a thousand heart specimens and imagined how the most flawed among them must have worked until their makeshift circulation systems gave up. Daily when she touched a patient, she felt the heart at work. On a screen, she watched it pump blood into the body, including along the alternative paths that could slow or divert blood. But like most other doctors in 1942, she had never actually seen an open chest or pumping heart.

Standing behind Blalock as he moved his scalpel, cutting into the chest, opening it to locate the duct, she saw the human heart beating inside for the first time. It furiously seized up and then let go, squeezing blood out and taking it in. "It was a great sight for me," she recounted. It was also momentarily terrifying. The anesthesia that should have slowed the heart's work during the operation was poor. They used open ether, an early method of dripping the drug through a piece of gauze on the patient's mouth. Toward the end, the child's heart was beating "awfully fast," Helen recalled. Blalock was also fast. He stitched closed the connections to the discarded duct in seconds. Then he closed the chest.

Helen thought his work was masterful, and she was waiting to tell him so as he exited the operating room. She began by saying, "Dr. Blalock, I stand in awe and admiration of your surgical skill," but then she continued, "The really great day will come when you *build* me a ductus for a child dying of anoxemia [oxygen shortage], and not [when you] tie off a ductus for a child who has a little too much blood going to the lungs." What he had just accomplished was not enough for her or, for that matter, for Blalock. This operation confirmed his skill on the operating table and showed Helen they could collaborate. His technique would prove better even than Gross's revised one, making it available to more children. In that room with Blalock, Helen saw the future and raced toward it. From the deflated look on his face, Helen knew that Blalock had expected congratulations, not another challenge.[9] He gave a great sigh. Whatever he thought, he agreed with Helen. It was one thing to cut off an errant blood path to the heart. It was quite another to reroute the blood to the heart itself.

"When that day comes," he told Helen, "this will seem like child's play."

10 CONVINCING BLALOCK

On a Friday morning in February 1943, Shirley and Leonard Rosenthal boarded a train for Helen's office in Baltimore with their nine-year-old daughter, Barbara. Their journey from Buffalo, New York, was fraught since the slight girl with wispy blond hair and beguiling brown eyes could walk no more than sixty feet without losing her breath and falling to the ground. Once inside the train car, her parents could have supported her, one on each side, to her seat or to the lavatory, but to get her to the train, Mr. Rosenthal very likely wheeled Barbara in a luggage cart.

From the day their daughter turned six months old and began to refuse food, they struggled to keep her alive. To sneak nutrients into her under-sized infant, Mrs. Rosenthal resorted to dripping a bottle over her mouth while she was sleeping.[1] She experimented with food after food until finally Barbara accepted bits of apple and chocolate. The search for nutrients for their daughter tested the Rosenthals to their core. When during a particularly desperate period the pediatrician suggested trying bacon to supply Barbara with much-needed protein, Mrs. Rosenthal learned to cook it from the Polish Catholic family in the downstairs apartment. Overnight, the family stopped keeping kosher.

Barbara had a heart malformation, tetralogy of Fallot, which deprived her body of adequate oxygen. It is the most common cyanotic heart defect; in the United States, about twelve hundred children are born with it every year. The main problem is an obstructed pulmonary artery that diminishes the volume of blood to the lungs, where it can be oxygenated. These children also have a hole in the wall between the two sides of the heart. Blood entering the right ventricle that should go to the lungs to be refreshed encounters resistance from the blocked pulmonary artery and is instead pushed into the left ventricle, where it mixes with oxygenated blood going out to the body. The blood to the body is further diluted by nonoxygenated blood from a misplaced aorta that straddles the right ventricle; normally the aorta receives only oxygenated blood from the left ventricle.

The shortage of oxygen in Barbara's blood interfered with the normal functioning of cells in her body. Some days she was so tired she could barely breathe. Nor would she eat. "Harrowing" is how Mrs. Rosenthal described Barbara's childhood, in which she and her pediatrician plied the child with food Barbara needed to grow just as she refused to take it. The trauma revealed itself in photographs: in some, Barbara appeared gaunt, and in others slightly plump.

Her father cheered her with a thick scrapbook in which they affixed colorful stamps from faraway lands like Ecuador and Montenegro. Her mother oversaw another sedentary activity, piano lessons. When Barbara worsened, Mrs. Rosenthal pulled her daughter in a wagon the block and a half to her school and helped her inside.

Day after day, this solemn girl with intensely blue face and lips sat at her desk, rarely speaking to conserve energy, occasionally dropping to the floor for relief, but listening intently and turning in homework, demonstrating over and over to the school's principal that "despite her severe handicap," Barbara was a "brave little girl with unusual mental capacity."[2]

On the Saturday morning in 1943 when Helen first met her, Barbara could barely climb half a flight of stairs or walk down the hall to her room before collapsing on her bed. Her mother would find her lying with her knees hugging her chest, panting, and unable to speak for half an hour. Barbara's squat was instinctive, a bid to push blood into the lungs and more oxygen into her body. Helen had observed even the smallest infant with tetralogy of Fallot bring knees to chest. The Fallot squat slowed blood going to the aorta so that it briefly reversed direction and was forced into the lungs, where it could be oxygenated. This extra oxygenated blood then circulated through the body, bringing relief just as the child was about to pass out. Afterward, the blood would return to its oxygen-diluted state until the child's next exertion forced her to squat again. Family outings to Lake Erie were anything but normal. In one photo, Barbara smiles sweetly as she squats in the sand. A second frame captures her distress: her eyebrows furrowed in a frown.

Left behind in Buffalo was the couple's one-year-old son, Paul. Mrs. Rosenthal had not wanted a second child, with Barbara so much work, but her doctor encouraged her because, he said, Barbara was going to die.

Indeed, Barbara was slowly suffocating. Her mother's attentiveness in the face of overwhelming evidence that Barbara was being asphyxiated had kept the girl alive. Mr. Rosenthal scoured library books and quizzed doctors until Derek Levy, a young cardiologist at the University of Buffalo, mentioned the rare doctor in Baltimore who studied children's hearts.

Helen examined the girl, recorded her blood pressure, listened to and touched her chest, felt the blood move through her arteries, and, after viewing her heart in the closet-sized fluoroscope room, confirmed the diagnosis: tetralogy of Fallot. It was an extreme case. Barbara's deep blue skin color signaled that her lungs were not getting enough blood to infuse with oxygen. Her soft, flat fingernails and toenails confirmed a severe obstruction in the pulmonary artery, the path to the lungs.

Over the next few days as Helen observed Barbara in the hospital, Helen explained the problem and very likely her idea of a surgical solution to the Rosenthals. With Barbara out of earshot, Helen advised the Rosenthals to treat the child as if they believed she was going to grow up. "It is unfair to treat a child any other way," Helen told her parents.[3]

This was consistent with what Helen had advised parents since the day a decade earlier when another girl about Barbara's age, the rheumatic fever patient whose family doctor had given up on her, had joyfully marched into her office on the rebound. That was the day Helen vowed to harness the power of hope and the willpower it summoned.

Helen could not have saved Dottie, whose weakened heart failed when her rheumatic fever later flared up again. But Barbara's heart was not weak from infection; in size, it appeared nearly normal. Helen knew what could save her: more oxygen.

More than three years had passed since Helen had approached Robert Gross in Boston to propose a surgically created blood vessel to improve oxygen in blue babies. Three months had passed since Helen's overture to Blalock as he exited the operating room. Her nudge had resulted in the equivalent of a shrug. *That will be the day.* Given Blalock's seeming indifference, the surgery probably wouldn't become a possibility in time for

Barbara, who was older than most children with the condition and in a constant state of respiratory trauma. Barbara was edging closer to the day when her body's oxygen demand would no longer be satisfied by squatting.

Helen picked up the phone and dialed Blalock's office and asked him to stop by when he was free. When Blalock arrived, she led him over to Barbara, who was hunched in misery near a door along the wall, to explain how her body struggled to feed itself oxygen. "That is the type of child you could help if you could build a ductus," she recalled telling him about Barbara.[4]

Stunned, the Rosenthals left Helen's clinic in an altered state of mind. No doctor had ever told them to treat their daughter as if she would grow up. For the first time, they learned of a possible solution to save Barbara. It would not have been lost on Mrs. Rosenthal that of all the doctors they consulted, only one refused to give up. With her own ears, Mrs. Rosenthal had heard the doctor lobby a surgeon on Barbara's behalf. A practical woman, she understood the limits of ambition in what she routinely called "a man's world." Yet the doctor taking this risk, pushing a solution, pestering this surgeon—that doctor was a woman.

The Rosenthals had given Helen a reason to push Blalock once more. In turn, Helen breathed new life into their decade-long battle. Hope boosted the Rosenthals when they most despaired. The next period of keeping Barbara alive would be even more challenging.

*

Barbara was not the only patient Helen introduced to Blalock in a concerted effort to pressure him to save her patients. Years later, Helen would laugh about this aspect of her strategy to convince him to build a passageway to the pulmonary artery.[5] Blalock was in and out of the children's hospital and Helen's clinic regularly as she and Park tried to interest him in the heart. She would point out an undeveloped body or a bluish tint around a nose or mouth of a child with tetralogy of Fallot; sometimes she held out a child's hand to show Blalock nails that appeared to float like silver moons on the tips of purple fingers. Park would recall on several occasions her pleas to Blalock for an operation to bring more blood to the lungs at a time when Blalock was thinking about ways to fix an obstructed aorta.[6]

But while Helen touted her idea to fix her blue babies in patient conferences, Blalock seized on a different project, one that Park had suggested in spring 1943: a blockage in the aorta, the body's main blood vessel (coarctation of the aorta). Quietly he took the problem to Vivien Thomas in the laboratory.

In children with this condition, the aorta is nearly cinched shut at the spot where the fetal duct once was, where the aorta turned downward to bring blood to the lower body. The aorta emerges from the left heart (ventricle) and moves upward toward the neck, where it arches and sprouts branches that carry blood to the upper body. Then it turns downward to bring blood to the stomach and other organs and, via more branches, into the legs. In a child with coarctation of the aorta, blood that was supposed to supply the lower body instead circulated to the upper body, leading to throbbing in heads and arms and, ultimately, death by heart attack or stroke. Park's patients were underdeveloped. To thrive, their aortas needed to be widened or cleared.

Blalock was fearful of removing the blockage altogether because if it was large, the remaining vessel parts might not be long enough to stitch back together. In the dog lab, Vivien Thomas could not even recreate it. After a patient conference that Park and Helen held for students and residents in a Harriet Lane auditorium, Blalock lingered to discuss the problem. This is where Helen made another pitch for her tetralogy of Fallot patients and finally captured Blalock's attention.

Their discussion revolved around making a bypass around the aorta obstruction. At the blackboard, Park sketched a branch of the aorta arch that takes blood to the upper body—either the subclavian that feeds the arm or the carotid artery that feeds the neck, accounts differ—that Blalock might reroute and fuse to the aorta after the obstruction as it descended. This would bring additional blood into the lower body.

Ultimately, these experiments would result in paralysis of dogs' legs, and Blalock hesitated to perform them. But that day, listening to the technical discussion, Helen seized on the fact that the section of aorta Park pointed to for the attachment was near the pulmonary artery, which took blood to the lungs. If Blalock could attach a blood vessel to the descending aorta to increase blood supply to the body, couldn't he sew one to the nearby pulmonary artery to deliver more blood to the lungs in her tetralogy of Fallot patients? This would increase oxygen throughout the body. "That's all I want," she told him.[7]

Helen realized that connecting a branch of the aortic arch to the pulmonary artery could produce that same workaround circulation as the open fetal duct, the heroic little passage directly connecting the aorta to the pulmonary artery that remained functioning, refusing to dry up when a baby experienced troubled breathing. A replica would reroute blood around the obstructed path to the lungs.

*

Helen's account of this meeting differs from those of the other participants. Park recalled that she clearly stated her patients needed a new path to bring more blood to the lungs but not a particular solution, "so that the suggestion must have been made at some time subsequently."[8] Blalock would say Helen never suggested the aorta-pulmonary artery link he made in her patients. "Concepts are grand, and I admire your excellent mind," he told Helen, but her "casual remark" that her patients could benefit from a means to bring more blood to their lungs was a far cry from developing and executing an operation.[9] It is clear, however, that Park's suggestion to try a different artery for the aorta experiment and Helen's comments that day about the need to deliver more blood to the pulmonary artery in tetralogy of Fallot patients struck Blalock as none of her previous pleas had. He realized he had already created a link like one she wanted in his laboratory.

Some years earlier in Nashville, in an experiment aimed at treating pulmonary hypertension, Blalock directed Vivien Thomas and a research associate, Sanford E. Leeds, to create a new passageway for blood in dogs by sewing the subclavian branch of the aorta to the pulmonary artery. The idea of creating an artificial shunt or blood vessel to get around obstructions or alter blood flow emerged before World War I. The Austrian surgeon Ernest Jeger in his 1913 book, *The Current State of Blood Vessel Surgery*, described his bypass between a vein and the aorta to improve blood flow—similar to the shunt Blalock asked his assistants to make—and a bypass to treat hypertension in veins bound for the liver.[10] Jeger demonstrated his technique on animals to the Hopkins surgeon William Halsted, whom Blalock revered, in Leipzig that year.

Blalock's link increased blood to the dog's lungs, though not enough to create the high pressure they needed to test novel surgical approaches on pulmonary hypertension. Could this artificially created vessel to bring blood to the lungs work with the complicated conditions in Helen's patients? This was a question for the laboratory. Within days of the conversation with

Blalock—immediately, Helen said—she found herself in Blalock's dog laboratory explaining her patients' problems to his assistant.

Blalock's invitation to Helen to visit his animal laboratory was unusual. Park had not been invited there to discuss the obstructed aorta. Blalock had not even mentioned his experiments until the results were summarized in his draft paper. He typically took ideas to Vivien Thomas, who executed them, the pair refining the idea in conversation, until they found something that worked. But Blalock knew little about Helen's patients or how their hearts worked.

He alerted Vivien Thomas to Helen's visit and walked the two blocks to the lab ahead of her, arriving early. The workaday space of the Hunterian Laboratory, named for the eighteenth-century Scottish surgeon John Hunter who famously collected animal anatomical specimens, had fans to disburse the heat and smell. It was full of caged dogs, metal examining tables, and operating equipment. Helen would not have winced at the sight of bandaged, lame, or traumatized dogs, though she might have reached through cages to pet them. Her fondness for dogs coexisted with her belief that experiments on them were justified if they saved human lives. From her teens, Helen supported and followed the public campaign by Cannon, the Harvard friend of her father's, to defend dog research by documenting human lives it saved.

In the lab, Blalock introduced Helen to Vivien Thomas. He would remember standing alongside Blalock when "this pleasant appearing, soft spoken, tall lady with a New England accent" began to talk. Blalock too listened carefully as Helen explained the problems of a child with tetralogy of Fallot. The defects associated with the condition occurred during fetal development and caused physiological changes in her patients. Different blood vessels took over for the usual ones, shifting blood flow patterns. As Helen had explained, a child with heart deformities remained alive only by what doctors called collateral circulation: some other vessel would take over to carry some blood, often expand a bit or push against walls, creating abnormal routes, some of which worked better than others and each of which left a distinct footprint or sound. The open fetal duct was one type of backup circulation. Knowledge about how such substitutes developed was limited, though doctors realized as early as the 1800s that compensatory routes could exist.

Helen also described the outward appearance of a child affected by a shortage of blood oxygen, which turned their lips, fingers, and toes bluer and bluer and forced them to the ground to hug themselves, gasping for breath. Her description concluded with her patients' death by asphyxiation. Vivien Thomas remembered that Helen seemed depressed when she told him that she watched every patient with this condition die. "She had followed these patients in the clinic and seen their condition gradually deteriorate until they finally succumbed."[11]

Two things stayed with Vivien Thomas about Helen: the overwhelming detail that she provided him on what was happening inside the bodies of children with this wretched condition and her complete confidence that children with these abnormalities could be saved. "She expressed her belief that, by surgical means, it should be possible to do something to get more blood to the lungs, as a plumber changes pipes around, but gave us no hint as to how this could be accomplished—what pipes to put where," he would recall. "She left us with the problem."

Blalock did not tell Helen he already had a potential solution, but Vivien Thomas immediately realized that he intended to try the subclavian-pulmonary shunt they had developed in Nashville in a failed attempt to create pulmonary hypertension. It was simple to install.

Recreating the abnormalities of blue babies in dogs, however, was a problem like no other. Vivien Thomas had never seen a malformed human heart. Repeatedly he visited the pathology laboratory where Helen stored her patients' hearts to study the abnormalities. "The reproduction of any of the conditions that existed in these congenitally defective hearts would surely not be easy," he concluded.[12]

For the next year or so, in the Hunterian laboratory, Vivien Thomas took time from his work on shock and attempted to mimic the defective circulation of a tetralogy of Fallot patient in dogs. The first challenge was to reduce the level of oxygen in the blood of a dog to that of a blue baby. He did this by stitching together part of the walls of the pulmonary artery, making it narrower, a condition called pulmonary stenosis. Less blood circulated to the lungs, but it was all oxygenated by the time it got to the body, unlike that in a tetralogy of Fallot patient, whose blood to the tissues was diluted by unoxygenated (venous) blood. Helen continued to discuss patients with Blalock and, by her account, to critique the animal experiments. Here is Helen's report of the effort:

[Blalock] tried to create pulmonary stenosis. This proved difficult, and as I pointed out to him, the creation of pulmonary stenosis would not cause cyanosis (bluish skin caused by lack of oxygen). So I suggested to his remarkable technician, Vivien Thomas, that he put the right pulmonary artery into the left auricle and thereby direct venous blood to the systemic circulation, and then when the dog became cyanotic, put the subclavian artery into the pulmonary artery to relieve this situation. Mr. Thomas performed the first stage of the operation the next day. It, however, took an additional operation on the lungs to produce cyanosis.[13]

In historical accounts of Blalock's work, Vivien Thomas said that Helen never appeared in his laboratory or contacted him after her first visit, indicating that perhaps she gave herself a larger role in the experimental phase of this surgery. He thought she gave the first detailed account of her patients' anatomy that day in his lab.[14] But he did not know when her discussions with Blalock began or how often they occurred.[15]

That Helen would have inquired about or critiqued animal experiments after her initial overture to Blalock in late 1942 fit her style. She had pushed for an operation, and patients who might benefit were on the brink of death. Helen assumed Blalock had begun at least thinking about her idea after the fetal duct operation.[16] That she followed up with him also seems evident from the fact that in February 1943, Helen revealed that something was underway. "We can't do anything now," she told the parents of Barbara Rosenthal, "but we're working on an operation and I hope maybe we can someday."[17]

The massive effort in Blalock's animal lab to test his shunt against Helen's theory continued well into 1944 and involved two hundred dogs. Vivien Thomas was able to create only two of the four defects associated with tetralogy of Fallot. Ultimately, he concluded he had wasted time trying. Nonetheless, in dozens of these dogs, he sewed a new blood vessel. It was like a construction project. Two arteries, called subclavian arteries, rise from the aortic arch to take blood to the right and left arms. For the operation on dogs, Vivien Thomas cut away one subclavian artery, temporarily clamping it closed to block the blood coming into it from the aorta, and then stitched it to an opening he made on the side of the pulmonary artery. Blood began flowing through the new path to the lungs as soon as he removed clamps. Thomas and a chemistry technician next studied changes to the blood, tracking changes in red blood cells and hemoglobin levels to assess oxygen levels. The dogs indeed got more oxygen, but the comparison to blue babies, with their additional abnormal conditions and makeshift circulatory routes, was imperfect.

*

Blalock was becoming increasingly discouraged over his prospects for surgical breakthroughs. On October 19, 1944, six months after he had published an account of his failed attempts to get around an obstruction of the aorta with a bypass, Swedish surgeon Clarence Crafoord performed the first successful repair, directly removing the obstruction. Gross in Boston would follow. And in the weeks following the aorta operation in Sweden, Blalock performed several major surgeries without great success, according to his chief resident at the time, William J. Longmire.[18] One patient had died, others were not healing as expected, and each contributed to Blalock's growing depression.[19]

While Blalock despaired, Helen developed evidence in her clinic to cement her theory that an ailing blue baby could recover if given a new blood vessel like the fetal duct. It had to do with a small boy and a new noise one day in early November 1944. While warming with her hands the stethoscope she intended to set on the chest of a 2 ½-year-old boy with tetralogy of Fallot, she noticed that the toddler was less cyanotic than she had remembered. Between visits, his skin had turned from blue to pink. This could only mean that more blood was circulating to his lungs to be oxygenized, and the boy's heart was pumping this good blood back into his body. How did it get there? As she bent over the boy, she heard a new sound: a continuous murmur that "sounds for all the world like a patent ductus arteriosus." She noted the second sound snapping in the middle and a harsh systolic murmur over the chest. It sounded to her that the boy's fetal duct had remained open after birth and somehow recently come into use, permitting more blood to flow through the lungs. This was astonishing; she had never seen a child with tetralogy of Fallot develop signs of an open fetal duct more than two years after birth. "Nothing could be better for him than to have his ductus open," she concluded. "Certainly it is going to improve his circulation to the lungs, which is what I am longing for Dr. Blalock to do for other small children."[20]

These were the circumstances in the second or third week of November 1944 when Blalock called Helen to discuss surgery on her tetralogy of Fallot patients. It was the only experimental procedure remotely ready that could bring a big return, and he was convinced he could easily install it, but he was not sure it would work. If the patient died, his career might be over. He needed encouragement. If Helen was convinced that creating a new passageway to the lungs in a blue baby would work, Blalock told her, he was willing to try it.[21]

Yes, Helen told him. It would work.

*

The decision made, Blalock called Vivien Thomas one morning to say he was coming to the laboratory to learn how to perform the procedure on dogs. Until now he had merely observed Vivien Thomas create the new vessel. To prepare himself to operate on a child, Blalock planned to assist Vivien Thomas several times before operating on a dog.

Helen reviewed her patient list. There was Eileen Saxon, a fifteen-month-old baby in intensive care. Born prematurely in August 1943, Eileen weighed five pounds when she left the hospital. At seven months, her lips and nails turned bluish, and she began screaming. Only after her frantic mother appeared in the emergency room did Helen diagnose a hole in the wall between the two bottom chambers of the heart (ventricle septum defect) and, later, an extreme case of tetralogy of Fallot. Eileen began to suffer cyanotic spells. For the past three months, she had lived in the oxygen chamber in the children's ward. For the past six weeks, she had refused most food. Although more than a year old, Eileen now weighed less than nine pounds. She was so fragile that the chief anesthesiologist refused to give her a drug for a minor procedure because he believed it would kill her.

The baby needed the operation badly, Helen told Blalock, "but I don't know if you want to operate on anything quite so small . . . and in poor condition."[22] Blalock was unfazed. A baby who could not live without the operation was precisely the candidate for this experiment, he told Helen.

Helen agreed to approach Eileen's parents. Her twenty-two-year-old mother, Dorothy Saxon, visited her daughter daily, walking a mile to the hospital from a modest brick row house on the edge of the city's Patterson Park. For months, she had cared for her daughter alone while her husband, Francis (Michael), a sheet metal worker, stayed late into the night building airplanes at the Glenn L. Martin Co. Now he was stationed in Norfolk, Virginia, with the Navy.

Mrs. Saxon, expecting bad news, froze when she learned Helen wanted to speak with her. When Helen sketched the aorta and explained how a never-before-tried operation might help Eileen breathe, the flustered young mother asked to discuss it with her husband.[23] The pair returned a few days later, when Helen introduced them to Blalock and again explained the risky operation. The Saxons agreed.

In the meantime, Eileen worsened. Fearing she would not live more than a few days, Blalock called Vivien Thomas early one morning and asked him to deliver the necessary tools to the operating room. Thomas raced to size and sharpen six needles and thread them with fine silk. He had to hold each needle, 1 1/8th inch long, in a clothespin-type clip to hone it down to less than ½ inch, small enough to suture the vessels end to side.[24] When vascular surgery took off, needles in every size and shape would be manufactured in bulk, but for this operation, Vivien Thomas worked long into the evening making each one.

Blalock next sparred with his chief anesthesiologist, Austin Lamont. Because the baby's oxygen levels were so low and one of Eileen's lungs would have to be compressed during the operation, Lamont questioned whether to move forward with the operation. He declined to participate; his first assistant agreed to step in for him.

The conditions for operating near the heart on the morning of Wednesday, November 29, 1944, were primitive, the standards in use on the brink of a paradigm shift. The participants lacked specialized tools and had little information about what they would find when they opened Eileen's chest. With no equipment for a baby so small, pain medicine for Eileen was administered by hand, ether mixed with oxygen and applied to a cloth over her mouth, rather than through a tube into her throat that would have allowed doctors to better regulate the amount of oxygen in the lungs.

Helen stood at Eileen's head. Blalock bent slightly over the tiny bundle on the operating table. Vivien Thomas, invited into the operating room for the first time, stood on a footstool, peering over Blalock's shoulder. He would talk Blalock through the operation.

The hand-sharpened needles Blalock used were smaller than half the length of his pinkie. Eileen's left subclavian artery was no bigger than a matchstick. Blalock cut away some veins and blindly reached into Eileen's chest and pulled out the tiny vessel. The pulmonary artery itself, the vessel taking blood to the lungs, was seriously compromised. Somehow, Blalock sewed together the miniature parts. Helen held her breath until she saw Blalock's worried look ease; at last, he felt the vibration of blood moving through the newly forged route.

Outside the operating room, where Eileen's mother had waited alone for hours, Mrs. Saxon heard her daughter crying as nurses pushed open the door to announce the operation was over. "If she's crying that loud she wants to live," she thought.[25]

The baby, slightly pinker in color, remained in critical condition. Trauma from the surgery caused her lungs to collapse. So began the vigil: night and day, a team of six doctors including Helen's fellow Ruth Whittemore took turns plying the tiniest of needles into the walls around Eileen's lungs to aspirate fluid, talking to her, and calming her. Mrs. Saxon visited daily but could not get around the doctors for two weeks. When she finally saw her baby, now with a more normal pink color, she felt as though she had witnessed a miracle. "I was beside myself with happiness," she remembered.[26]

Helen was already imagining other patients with skin as pink as Eileen's. "She gave us courage to continue," she would say.[27]

*

News of Eileen's successful operation was closely held, but Helen shared it with Marjorie in Oxford, who was celebrating the publication of her book on fetal circulation. Thrilled, Marjorie asked for more details on the sly; she and her colleagues had dissected so many fetal and postnatal hearts that they were keenly interested in the surgeon's technique.[28]

Helen knew Blalock wanted to wait a month to see how Eileen fared before trying the experimental procedure again, but even on the day of Eileen's operation, she began to assess potential future patients. In her clinic hours later, Helen reexamined Marvin Mason, a six-year-old boy with a severe case of cyanosis, and decided he would be a good candidate for the second surgery. Marvin could no longer walk, and his skin was almost navy, but he was younger than Barbara Rosenthal and had fewer complications. He would be more likely to survive an operation, and success was critical given Blalock's caution. Helen gently mentioned the possibility of surgery to prepare the mother, shocking her.

Within two weeks, after Eileen Saxon's lungs had cleared, Helen sent her social worker across the city to the Masons' house to ask Ruth Mason if she was interested in the operation for her son. Yes, if Helen could get her husband released from the Army. Marvin Mason's father coincidently arrived home in December, and when the family returned to see Helen, she scheduled Marvin for the second blue baby operation for February 3, 1945.

She had not forgotten Barbara Rosenthal. Now eleven years old, she was losing consciousness for up to half an hour at a time. By phone with Barbara's cardiologist in Buffalo, Helen related news of the historic operation. She suggested Barbara return for an exam in January.

Helen took the train to Boston to spend Christmas with her sisters and their families. The end of World War II had seemed near. The Allies were celebrating a liberated France. All summer, as Helen anxiously awaited the results of Blalock's laboratory experiments, Americans mourned the deaths of thousands of American, British, and Canadian troops who crossed from England to France and in June 1944 scaled the beaches at Normandy in a herculean effort to oust the Germans. By the time German troops left Paris in August, tens of thousands more, many of them civilians, suffered or died in Allied bombings. Deeply disturbing reports before Christmas told of mass executions by the Germans, who were engaged in a fierce but futile last offensive. Throughout the war, hundreds of thousands of young men had been intentionally killed or injured in battles too horrific to imagine; to Helen, who was working every day to give children maimed at birth through no fault of their own a chance at life, the dissonance would have been extreme.

When Helen returned to Baltimore in January 1945, she found little Marvin Mason consumed by a respiratory illness and sent him home to recover. When she reexamined Barbara on Saturday, January 6, 1945, she noticed that even getting from her wheelchair to the examining table left Barbara breathless. She could walk half as far as when Helen first observed her nearly two years earlier. With Blalock's approval and Barbara's consent, Helen slotted Barbara in for the operation on February 3.

The Rosenthals made preparations to be in Baltimore a month or more. Mrs. Rosenthal arranged for Paul, now almost four, to live with cousins and to enroll in the school they attended. At Barbara's request, her parents took her to the theater in New York City. Mrs. Rosenthal engaged Barbara in redecorating the girl's bedroom. "It's going to be nice. And you'll be well and wouldn't it be nice to have what you want in your room?"[29] Helen

remembered Mrs. Rosenthal recounting to her. The truth was, Mrs. Rosenthal did not believe her daughter would survive. The new wallpaper, drapes,
and furniture were a preemptive bid to dispel memories and assuage what
she anticipated would be her own terrible sorrow when she arrived home
without Barbara.

Mr. Rosenthal was more optimistic. When the stationmaster asked
whether the three train tickets he intended to purchase to Baltimore were
to be one-way or round-trip, he did not hesitate. They were all round-trip.

11 THE DAWN OF MODERN HEART SURGERY

On January 25, 1945, Helen discharged Eileen Saxon, who had survived her first difficult weeks and after two months in the hospital finally gained weight. Because she was so frail, doctors decided against poking her with needles to test her progress. In place of blood tests, Helen assessed Eileen with data she could see with her own eyes: the child's pink skin, her appetite, and her ready smile. For Barbara, Blalock suggested a more objective way to measure improvement—the change in oxygen saturation in the blood or how much oxygen these blood cells carried after surgery.

Helen expected the oxygen level of a child in respiratory arrest to be low, but the preoperative test by Vivien Thomas's chemist shocked her. Barbara had only about one-third of the oxygen normally required for life. For the first time, Helen and Blalock understood how little oxygen patients with tetralogy of Fallot lived on in the final stages of their lives.

This data highlighted the operation's alarmingly narrow margin for success. A patient already starved for oxygen would undergo an operation that required shutting down one of the two pulmonary arteries that brought blood to the lungs. How to supply oxygen to the patient during the operation was now the most critical problem. Blalock had considered it for a year, hoping for a method to provide it directly into her bloodstream as if she was breathing on her own, because she would be immobilized. Instead, Barbara would receive oxygen, mixed with anesthesia, via a tube placed down her throat.

Austin Lamont, who had refused to participate in Eileen Saxon's operation, agreed to administer cyclopropane mixed with a high concentration of oxygen. This relatively new analgesic agent worked quickly and smelled good, though it would eventually be banned—combined with too much oxygen, it was combustible.

The operation began at 9:00 a.m. on Saturday, February 3, 1945. As soon as Blalock opened Barbara's chest on the left side, he knew something was wrong. The vessel he expected to cut, turn downward, and stitch to the pulmonary artery was not there.

Helen's preoperative report noted that Barbara's heart had an unusual structure. Her aorta arched to the right. With the position of the aorta reversed, the subclavian artery Blalock hoped to attach to the left pulmonary artery was on the other side of the arch and too far away. He could not even find it.

Normally, three vessels come off the aorta as it arches to the left to bring blood to the upper body: the first and biggest, the innominate (also called brachiocephalic), branches after a few inches into two smaller arteries that feed the right arm (right subclavian) and the right side of the neck and head (right carotid). The next two vessels spring directly from the aorta to feed the left side of the brain and left arm (the left carotid and the left subclavian.)

Oddities abounded inside Barbara's chest. Blalock found a large vessel running down the left lung, an extra bronchial artery that appeared to supply some blood to the lung. Other vessels were larger than normal; they too were picking up the slack, sending blood to the bottom of the lungs. Helen and Blalock had presumed Barbara had some unusual circulation inside her chest; otherwise, she would be dead. But the sight of this mass of intertwined smaller arteries greatly disturbed Blalock. For a moment, he stopped to discuss whether to continue the operation.

Could they stitch a different artery in place of the obscured subclavian artery? The most accessible artery was the innominate (brachiocephalic) artery, which rose a few inches off the aortic arch and split into two small arteries. To use it, Blalock would have to cut off branches that took blood to the left arm and the left side of the brain. If Barbara survived, the remaining carotid artery, the one feeding the right side of her brain, would have to expand its mission or she might get a stroke. This was not the method that Vivien Thomas had rigorously tested in the laboratory, but Barbara would die if he did not try it, so Blalock proceeded.

A large part of the two hours and forty minutes for this operation was taken up discerning the route of the makeshift set of blood vessels that had been keeping Barbara alive and finding and separating the innominate vessel from surrounding tissues. Once he separated the innominate artery, he clamped it in two places to prevent any blood from escaping when he cut and tied off the two branches sprouting from it, the carotid to the brain and the subclavian to the left arm. He also clamped shut the left pulmonary on either side of the site where he planned to attach it so that blood could not escape while he stitched; this key artery (one of two) taking blood to the lungs would be shut for at least fifty minutes. Remarkably, Barbara's blue color did not deepen. With the clamps in position, he began stitching the end of the innominate artery to the side of the left pulmonary artery.

Helen watched Blalock's hands shift and fingers fly. She could not hear him remove the clamps, but she saw a sudden flurry of activity around Blalock as he called for a needle again. A leak sprang up as blood rushed through the new passageway; quickly he stitched it shut, then tested the pulmonary artery. He felt the vibration of blood pushing on the artery walls, the "thrill" that accompanies fast-moving blood from the aorta into a new space with less pressure, as it moved past the link he had created. A substantial new path now brought more blood to the lungs in hopes of increasing oxygen to the body; before their eyes, the pulmonary artery taking on this extra blood expanded. Blalock reached for a bandage. Barbara stirred.

Helen watched Barbara, hoping to see the same transformation she had witnessed in Eileen, but there was no change in the girl's skin color, and Barbara's left arm had no pulse. Still, at Blalock's request, Barbara lifted her arm skyward. Her response showed the operation had not affected her brain.

It was now clear that patients could survive for up to an hour with one of two arteries that normally take blood to the lungs blocked. This was extraordinary; it meant that the pulmonary artery and lung on the opposite side of the chest had been doing twice as much work all along, or else blood was trickling into the lungs via other arteries that had taken on the job of the pulmonary artery. In Barbara's case, the huge bronchial vessel and multiple alternate routes led Blalock to conclude that the valve on the pulmonary artery, which normally opened and closed to allow blood from the heart's right ventricle to flow into the artery and on to the lungs, had never worked. Instead, a swirl of blood vessels had adapted to push blood to the lungs as best they could. The process must have been gradual; she would not have been able to live from birth with a closed valve.

Probably her fetal duct remained open long enough after birth for some of this alternative circulation to develop. Blalock would use the word *tortuous* to describe the large and winding vessel that had grown inside Barbara to keep her alive.

The Rosenthals saw their daughter immediately after surgery, and they thought she looked better. Helen agreed but urged caution. "Let's wait and see," she told them.[1] The girl was breathing inside an oxygen tent, which might explain her slightly improved color. The thrill Blalock felt signaled that blood was flowing in the right direction, Helen knew. But would it pick up more oxygen in the lungs? She and Blalock had different theories of what might be happening. Blalock thought it was possible that Barbara's lungs, which had long operated with reduced blood flow, would have lost some of their function and be unable to manage the bigger volume of blood. Helen feared Barbara might have clots along the path to the lungs because of her high red blood cell count. That would make it harder for her lungs to expand and begin oxygenizing the onslaught of extra blood.

Helen made sure penicillin was running through Barbara's bloodstream to prevent sepsis infection as she was whisked to the recovery room. Because of the war, this powerful new antibacterial drug was in short supply, rationed for extreme cases. Helen had experimented with it briefly in her clinic hoping to prevent a recurrence of rheumatic fever. Only that morning its promise was heralded in *JAMA*, which reported successes using penicillin to treat meningitis, syphilis, and strep throat.

Also in the news that morning, American soldiers in the Philippines opened camps there to free four thousand islanders imprisoned by the Japanese. World War II was surely in its final months. Closer to home, an event in downtown Baltimore made Helen realize that societal breakthroughs that allowed Black men like Amy Clark's husband, Lee Clark, a military veteran, to work side by side with White men at Fort Holabird, would be difficult to sustain after the war without persistence. She felt strongly that Black men should have opportunities to succeed outside the military.

It was a parade, one step in a subtle but ongoing campaign for equal treatment from department stores. The marchers on Howard Street carried posters that rated stores by their service to Black customers. Hochschild Kohn's, with a no-returns policy only on clothes purchased by Blacks, was marked with onions. Hutzler's, one of the first to welcome Blacks, got orchids.

*

On the third day after the operation, Helen heard an extraordinary sound in Barbara's heart: the continuous murmur she had expected immediately after the operation. This was the sound of blood moving under high pressure through the new blood vessel into the slower pulmonary arteries on both sides of her chest. An X-ray picture would reveal Barbara's heart had expanded to accommodate the oxygenated blood now flowing there from the lungs. Helen added a blood thinner, double dose, to dissolve any clots in her lungs. Barbara's red blood cells were dropping. Her lips now appeared slightly reddish.

The third operation was a few days away. Blalock began to think that the artery he had unexpectedly used for Barbara might be a good fit for other extremely sick patients too. Larger and sturdier than the subclavian, it could move more blood, and it was closer to the pulmonary artery. He discussed it with Helen and Park, and they agreed that Blalock would try to choose which vessel to redeploy based on its size and the patient's need for oxygen.[2] He would select a smaller vessel for a patient whose impairment was not so grave, for example, since a bigger one might overwhelm the lungs. These decisions would have to be made before the operation, so he would know which side of the chest to open.

The third patient, six-year-old Marvin Mason, was as oxygen deprived as Barbara, and as a result, Blalock decided to try the bigger artery. Emaciated, Marvin weighted thirty-four and a half pounds, two-thirds of what a child his height and age should weigh. Helen described the membranes of his lips as mulberry, the color of the blackish-purple berries on trees on Cape Cod. Whiny and dour, Marvin refused to walk even a step for Helen.

One week after Barbara's operation, on February 10, 1945, Blalock cut into the right side of Marvin's chest, where he found the innominate vessel as expected in a normal aortic arch and prepared to attach it to the right pulmonary artery. He cut and tied off its branches—the arteries taking blood to the arm and head on that side—and began to separate the innominate vessel from surrounding tissue. It was harder to see the whole vessel on this side of the chest, however. Two clamps, one on each side of the section where Blalock had cut the artery, could not hold shut the larger vessel, and he was forced to attach a third to stop the bleeding. Then, when Blalock released the clamps to open the new passageway, he realized he had missed stitching a section. Suddenly Helen saw a terrible hemorrhage. "Blood just welled up," she said.[3] Instantly, the boy's blood pressure dropped. The linens on the operating table turned red and the little boy turned "desperately" blue. Helen was sure the end was near for her patient. "My heart sank," she remembered.[4]

By the time Blalock reapplied clamps to stop the external hemorrhaging, the boy had lost nearly 10 percent of his blood. Blalock was surprisingly nervous around big bursts of blood. He had planned for this moment by storing plasma in the operating room. An expert on preventing deaths from blood loss, his recommended techniques were in use that very Saturday halfway around the world to save American soldiers. As Blalock reapplied the clamps, his assistants simultaneously pumped the boy with plasma, making sure all those red blood cells trying to carry oxygen continued to flow to other parts of the body.

Amid the chaos, Helen emerged as what she called "the stabilizer" of the room, keeping others calm to prevent them from distracting Blalock, who upset easily. When the young anesthesiologist, Merel Harmel, frantic at the boy's suddenly blue appearance, asked if this operation was a do-or-die situation, Helen shushed him. "Just be quiet," she said brusquely. "It will be alright."[5]

Blalock located the leaky opening and repaired it with a "mattress" suture, a type of stitch useful for huge gashes and flimsy, difficult-to-attach skin. He would later ask Vivien Thomas to design a new tool to improve his reach to this artery.

Blalock next released the clamps and pressed his fingers on the pulmonary artery. He felt a "beautiful" thrill, according to Helen. The anesthesiologist called excitedly, "He's a lovely color now!"[6] The child who an hour earlier had been nearly purple and close to death now had bright pink cheeks. The color of his lips reminded Helen of the cherries that birds plucked from bushes outside her Cape Cod cottage: red.

His personality changed abruptly too within minutes of a bigger supply of oxygen creeping into his bloodstream. Awaking from light anesthesia as Blalock bandaged his chest, Marvin began talking politely, as if years of suffering and the rerouting of his circulation system an hour earlier had been but a bad dream.

"Is the operation over?'" the boy asked.

Blalock said, "Yes."

Immediately, according to Helen, the boy asked, "May I get up now?"

Blalock advised him to "please be still a little longer."

"Well, may I have a drink of water?"

From a boy who refused to eat or walk, a boy who snarled at any task that required energy, still on the operating table, such words were astonishing and gratifying and conclusive. A quiet descended on the room, Helen recalled.

"We knew then and there, we'd won."[7]

Four days after the operation, they removed him from oxygen. Marvin recovered so quickly that Helen engaged in a tug of war to keep him in bed, enlisting nurses to find more toys so Marvin could entertain himself. For three weeks, the boy pestered Helen until she allowed him to sit up, to stretch, and eventually to put his feet on the ground. He had to learn to walk again, and she wouldn't let him try without someone else present. If Marvin experienced pain during his recovery, he did not complain. From dour and snarling, he transformed into a happy and sunny child anxious to get up.

*

This was how the era of modern heart surgery opened, quietly, after centuries of theories and decades of trials and perseverance. In those early months, only the Hopkins staff and the referring doctor in Buffalo knew history was being made. Until now surgery on the chest had been limited to removing obstructions or closing stab wounds; these three surgeries to reorder vessels into the heart would pave the way for surgery on blood vessels and ultimately to actual repairs on the heart itself.

Barbara's recovery, however, was still a question mark. Kindly nurses allowed her parents to stay at her side long past visiting hours. Helen or her assistant visited several times daily, taking note of Barbara's improving color. It was not until two and a half weeks after Barbara's operation—a week and a half after Marvin's dramatic transformation—that Helen decided it had worked and allowed Barbara to stand and take a few steps. A few days later, Helen gave the go-ahead for Barbara to walk. Mr. Rosenthal positioned himself at the end of the hallway on her floor. Barbara walked a full sixty feet to her father, rested briefly, and returned to her room. We can only imagine the man's smile. The normally loquacious Mrs. Rosenthal would describe Barbara's recovery as "a miracle beyond words."

A few days after Barbara's hallway triumph, Helen learned that her blood oxygen saturation level measured an extraordinary 82.8 percent compared with a preoperative level of 36.3 percent. Barbara's blood would never be 100 percent oxygenated, not with her defective heart still allowing some nonoxygenated blood to mix with the additional oxygenated blood pumped out to the body. But it was more than enough to prevent cyanosis. Her fingers and toes, so long starved for oxygen, began to return to their normal shape.

The family stayed in Baltimore six weeks while Barbara recovered at St. Gabriel's convalescent home, including an extra week at Helen's request that allowed Barbara to participate in the public unveiling of the operation.[8]

On March 12, 1945, Helen and Blalock paraded Barbara, Eileen, and Marvin before an audience of Hopkins doctors, staff, and medical school alumni. With the children arranged so everyone could see their pink skin and healing fingers, Helen and Blalock gave the first public description of their breakthrough operation. The children demonstrated how they could now walk (Eileen was just learning). Helen gave each child a stethoscope to hold so doctors could step up to the patients to hear for themselves the continuous murmur that signaled expanded blood flow to the lungs.[9]

Park had come to Helen's clinic to congratulate her after the first operation. Now he told others that in Blalock, Helen had finally found her "daring young man on the flying trapeze," a reference to a popular song recorded by Rudy Vallee. This was public recognition of Helen's crusade to unravel the cause of her patients' deaths, discern a solution, and convince a surgeon to go into the abyss to test it.

And it was recognition of the great challenge Blalock took up. Blalock would be celebrated for a surgical feat like no other, an operation to change the way the body worked. This primitive excursion into the heart's workings revealed that patients with birth defects had ancillary backup systems that stood them well during a surgical intrusion that on the perfectly formed patient would be fatal. The trapeze he walked was flimsier than suspected. Helen had obtained proof that her theory would work when she detected the sound of extra blood flow to the lungs and improved color in a two-year-old. But the diagnosis she made days before she convinced Blalock to operate was wrong. Instead of a fetal duct that suddenly started working, the boy had a truncus arteriosus, a single large vessel in place of a pulmonary artery and aorta, which was confirmed when he died twenty years later. Helen had previously diagnosed the condition in only three infants. She would later describe how to distinguish the vessels.

Blalock's brilliance as a researcher set the stage for his success. For Helen, the success of her idea on the operating table was the culmination of a decades-long campaign she staged on behalf of her patients. The operation itself was a technical issue, a tactic in a general's strategy in war, hence her use of the term, "we'd won." Their partnership so cemented, Helen would speak of her work with Blalock collaboratively, using the term *we*. Each knew they could not have moved science forward without the other. But Helen's role in the beginning of heart surgery would be eternally second-guessed and her credentials as a scientist, one who seeks understanding of problems and searches for answers, often dismissed.

In spring 1945, Blalock was elected to the National Academy of Sciences. It was a momentous honor, rare for a surgeon. Nomination letters and those who write them remain confidential almost three-quarters of a century later. Given the timing of the announcement, the same spring as the *JAMA* article on his breakthrough surgery with Helen, it is reasonable to assume that Blalock's election was predicated on his 1930s work to clarify the nature of shock from battle wounds, although his work with Helen was known to associates months before the announcement. Blalock's analysis of traumatic shock and his method of administering plasma to counteract reduced blood volume was a fundamental step forward for medicine. It made surgery safer. Blalock would say it was his most important work.

Helen's congratulations and her joy when Blalock was fêted were genuine; after all, his research had helped save her patients. Still, coming as the acclaim did just as the pair won worldwide recognition of their dual accomplishment to introduce heart surgery, she would admit she felt hurt. She had isolated the mechanisms of defective hearts and suggested a solution to alter human circulation. Her idea was transformative. She got neither a promotion nor a pay raise.

*

The historic passageway Blalock installed in blue babies is known as the Blalock-Taussig shunt. (A more appropriate label would be the Blalock-Taussig-Thomas shunt.) The first three lifesaving blue baby procedures were referred to or posted by hospital staff as the Taussig-Blalock operation.[10] Soon after Helen and Blalock unveiled the operation to hospital employees in early spring 1945, Blalock became intensely jealous of Helen's billing, and according to several accounts, he tried unsuccessfully to remove it. Warfield M. Firor, a Hopkins surgeon who temporarily chaired the department before Blalock's arrival, recalled that Blalock "did everything to eliminate her name and everybody in our department knew it."[11] He remembered a tense atmosphere. For a while, he recalled, "I am sure Al Blalock didn't speak to Helen Taussig."[12]

Park mediated. "The only time I have seen Ned Park really angry was when he told me about this and about the lengths to which he had to go in order to ensure that Blalock gave Taussig proper credit," recounted Lamont, the anesthesiologist for early blue baby operations.[13]

At the time of their spellbinding talk to hospital staff, Helen and Blalock were finishing the *JAMA* paper that would publicly unveil the operation. As they exchanged drafts, Helen became convinced that her contributions outweighed Blalock's and told Blalock her name should come first. She had identified the problem and the theoretical solution. The patients were hers. She had diagnosed them, figured out a solution for them, and convinced Blalock to try her idea of a surgical rerouting to bring more blood to the lungs. Her demand was audacious; no one had first billing over Blalock on a paper about a surgery and certainly not one of this magnitude. The trail of memos is cold at this juncture, but that there was rancor and negotiations for credit is clear. Helen pressed her case repeatedly with Blalock.

Finally, she devised a way to identify her contributions. Soon she handed Blalock a version of their paper with her initials sprinkled throughout in parentheses (H.B.T.) to denote her original work. This unusual tagging of individual contributions at critical junctures revealed how much of the material in the paper was hers.

The first name on an academic paper belongs to the person with the original idea. Typically that would be the professor, followed by his or her researchers in order of their contributions. Helen was not a Blalock protégé; she was chief of her own clinic and had published multiple articles under her name. They were two people from different parts of the hospital. Helen had crossed over into surgery to find a collaborator for her idea to help children.

There was no precedent for this collaboration between branches of surgery and medicine. Few could even imagine it; specialization within medicine was only beginning. The operation prompted the then young American Heart Association to form a subgroup for doctors interested in vascular surgery. Helen herself would develop the study of children's hearts into a key division of medicine. She was showing Blalock what could be done for children whose problems he knew little about and where to apply surgical techniques to greatest effect. His idea was the how of it—what precisely to develop in his laboratory to connect the two parts of the circulation to direct more blood to the lungs. The implication of their relationship was enormous. The draft paper made it sound as if he had worked with their patients all along to discover what they needed. That rankled Helen. Also upsetting to her was his attempt to take credit for outcomes; afterward, both had cared for these children.

One explanation for Blalock's behavior is that he believed the most important advance described in the paper was the surgical procedure itself, which was indeed transformational in humans and followed remarkable work in his animal laboratory years earlier in Nashville as well as in Baltimore. His hand was on the knife. And he may not have known about or valued as equal to his experimental animal studies the decade Helen spent in her clinic observing and analyzing clues from thousands of children to learn what they needed.

Alternatively, perhaps Blalock considered it unthinkable to share credit for this historic advance in surgery with anyone and, at that, someone he considered of lesser stature. He was the chairman of a department with a storied history; she was an instructor in a young and unglamorous specialty— pediatrics—often populated by women, and although she headed her own clinic, few besides Park and her patients knew why it existed. He generously shared credit when he benefited by association. Over Park's protest, he added Park's name to the coarctation paper when Park had done little more than make suggestions. In Nashville, he gave top billing to a research associate, Sanford E. Leeds, on their joint paper describing the new blood vessel that would become the Blalock-Taussig shunt because Leeds, a fellow doctor and promising researcher, had made it.[14] Vivien Thomas, equally responsible for the laboratory work, was not named. He was an unknown "colored" technician. Nor did Blalock credit Vivien Thomas for the dog experiments behind his theory of shock—experiments that made Blalock famous. Like

Blalock, Vivien Thomas had no formal training in vascular surgery: he had taught himself beginning with techniques described in books.

Park did not agree with Helen that her name should be first. To him, the advance described in the paper was a surgical procedure on a human being; if it failed, Blalock would take the blame. "I had no sympathy for her point of view, and told her so," he would recall.[15] Knowing Helen's magnanimous nature, he found it very difficult at the time to understand her demand. In a compromise, Helen removed her initials from all but two sentences, and she was hopeful when Blalock told her he expected her name would come first on a future paper of theirs on the long-term success of patients. But she was not appeased.

Helen would never be satisfied with how her contribution was portrayed, habitually revisiting it, and would continue to insinuate herself into surgery on behalf of her patients. Her effort ensured that collaboration would become the model for patient care even as surgeons were ascendant, undergoing rigorous training and shaking the disdain with which they had been held for centuries by classically trained physicians like Helen.

Blalock would continue to try to separate Helen's work and name from his papers. Unable to sever himself completely because he depended on Helen for her diagnoses and patients, he deployed surrogates. Some of the brightest young surgeons around him would not learn for decades the role Helen—or Vivien Thomas—played in the beginning of modern heart surgery.[16]

On May 19, 1945, the article by Helen and Blalock describing the success of the Blalock-Taussig shunt on the first three blue babies was published in *JAMA* with Blalock listed as the lead author and Helen's name listed second.[17] Reading it today, more than seventy years later, one still senses the enormity of the change it described and the shift in the understanding of medicine's potential to help desperately ill people. The dawn of therapeutic heart surgery quickly became front-page news everywhere, except in Baltimore. The *Baltimore Sun* had managed to break news of the operation two days earlier; the story quoted Helen extensively, highlighting the fact that one of the doctors behind the surgery was a woman.

The newspaper, in fact, had reported the surgery five weeks earlier, buried in a story about dogs. At a boisterous three-hour city council hearing on a bill to ban dog experiments, replete with examples of cruelty in the laboratory, Blalock and a string of prominent doctors told of the lifesaving blue baby operation and other discoveries possible only with dog surgery. Then, in a tableau arranged by Helen, children with a once-fatal congenital heart condition walked happily into city hall—Barbara Rosenthal and Marvin Mason, with their teary-eyed fathers—followed by an Anglican nun in full habit from St. Gabriel's. The antivivisectionist movement sweeping the country had won victory after victory, but the Baltimore City Council decided the health of the citizens was at stake. The ban was defeated.[18]

12 PAPER FIGHT

The universal acclaim in academic circles over the *JAMA* paper cooled their fire over the summer, but there was a subtle fraying of the Blalock-Taussig relationship as each tried to figure out how to move forward with, or without, the other. It was a dance as delicate as a surgery. Blalock sought counsel from Park to justify omitting Helen's name on some publications. Helen, anxious to preserve a collaboration that would benefit her patients, continually reassured Blalock both that he was brilliant and that his success depended on her. When Blalock worried that one or the other of them was out of town too frequently, missing opportunities to operate, she cautioned against working too hard and reassured him that her upcoming book and invitations to speak would bring them more patients than they could handle. "When my book comes out it will give us more than a handsome edge in the field," she told him. "The real vindication of the work is that it will be accepted & others will learn to do it. Those who can afford it will still want the best & they'll come to you."[1]

Mindful of Park's criticism, she would tell Blalock that she need not be included on all surgery papers, such as ones explaining the mechanics of closing a fetal duct. But she expected to coauthor solutions to heart defects she had first diagnosed and explained, and she fought to publish collaboratively what she and Blalock learned from these new surgeries. She even plotted out a series of jointly written papers, and for each topic, she told Blalock, the one more interested or knowledgeable would take the lead. One paper she was drafting focused on the volume of blood that reaches the lungs and would compare the specific arteries surgically rerouted by

Blalock to postoperative improvements in blood oxygen. Another paper would examine how heart size changed after an operation. She listed Blalock's name second on these papers.

*

In late July 1945, Helen escaped to Cotuit for five weeks, instructing her senior fellow Whittemore to phone her daily to discuss patients Blalock would operate on before leaving for his own vacation. She had hardly settled in to enjoy the cool breeze on the beach in front of her cottage when she began to worry that Blalock would choose the wrong size artery for one little boy. Her concern followed a recent operation on a young girl in which Blalock had expanded the blood supply using the innominate artery, only to find afterward that the patient's heart was dangerously enlarged. The boy to be operated on did not need that much extra blood to his lungs. His heart was already large and his heart action "a little more tumultuous" than that of another patient. "I trust you will forgive my telling you what is in the back of my mind," Helen wrote Blalock, asking him to use the boy's subclavian artery, assuming it was fair-sized, which she was quite sure it would be. "It would be safer . . . even though it might be necessary to do a second operation in subsequent years."[2]

Helen tried to be gracious when she proffered advice, voiced an opinion, or made a demand to Blalock about his operations, but she could never fully entrust her patients to him, not when she was the authority on the unusual workings of their ill-formed bodies. Although her deputy could relay her wishes to Blalock, she still wrote to him to reinforce her views. In the beginning, he hadn't minded. At Helen's suggestion, he had even begun installing their shunt into children with atypical, extremely difficult-to-diagnose conditions like those in tetralogy of Fallot blue babies. When he reported a week later that the boy she worried about seemed fine, Helen expressed delight and thanked Blalock for using the smaller artery.

So far she considered his results "extraordinarily brilliant," she wrote to the director of the Commonwealth Fund in late July in search of a three-year grant. The pair were "head-over-heels" in the surgical treatment of children with narrow or completely blocked paths to the lungs (pulmonary stenosis and atresia). Of the thirty-four children Blalock had operated on, twenty-four were markedly improved. Seven had died.[3] This was a death rate of 20 percent, enormous today but at the time quite acceptable for a novel surgery on children destined to die within months or a few years. The grant would help her follow these patients.

Helen was in Cotuit when she learned of the death of Eileen Saxon, the pioneering blue baby. The toddler had become a local sensation, courted by photographers as she played with her Pekinese dog and visited a park with her father on his weekend leave from the Navy. But the matchstick-sized blood vessel did not provide enough oxygen for long. Her breathing difficulties returned. In an emergency operation, Blalock entered the other side of her chest, hoping to open a new passage to bring blood to her lungs. But Eileen was too frail and that blood vessel too was not big enough. She lived only five more days.

Eileen's death was not unexpected, but deaths had been mounting since the spring, and some were inexplicable. "The news is mainly bad," Blalock wrote Helen in August after three more deaths, including that of a boy he believed would do well. The boy developed dangerous air pockets in his lungs and chest. Blalock did not understand why, and it greatly upset him.[4] He admitted to Helen that he was "rather low in my mind."

Helen both sympathized with Blalock and obsessed over these patients' outcomes, comparing one to another. In reply to Blalock, Helen offered ideas about what happened, asking him whether it would have been better to suture the ends of two vessels together rather than an end-to-side, for example, in a case where the pressure in the pulmonary vessel was higher than usual and the blood flow small. She could visualize the circulation of the child's malformed heart, even if she couldn't explain how to repro-duce it in dogs.[5] She freely admitted knowing nothing about operative tech-niques, relying on Blalock to make his own judgments on the operating table. But Blalock had everything to learn from Helen about her patients' conditions and how their bodies worked.

Ensconced in Cotuit, Helen had every reason to be optimistic. The defeat of Germany and surrender of Japan signaled a return to normalcy and a new world order, with America at the forefront. She was drafting a new end-ing to her book unraveling puzzles of the heart—a chapter describing the first surgical solution for babies with defects. On her beach one particularly fine afternoon, she relished watching her nephew reel in seventeen butter-fish, "which is a pretty good record for a nine-year-old," she wrote Blalock.[6]

She ended her letter to Blalock with a wish that he too enjoy a grand vacation. "Your stage is certainly well set for it—after a spring such as you've had & now with the war over," she wrote. Her sentiment reflected on her own similar experience of the previous few months and her anticipation for their future work together. Everything had changed; her work was finally taking off. They had the chance to help so many children.

"Isn't it wonderful?" she asked. "What could be better?"[7]

*

By the end of the summer, both were writing papers on aspects of the shunt Blalock had created for tetralogy of Fallot patients. Seeing an opportunity to clarify her role in the original operation, Helen included a paragraph stating it was her idea to build an artificial passage to bring more blood into her patients' lungs. She sent it to Blalock—they routinely exchanged drafts—and now it was Blalock's turn to feel dismissed.

"In the paper which you have just sent to me you claim credit for the entire concept, but you do not state that I performed the animal experiments, that I devised the method, that I performed the operation on patients, that I suggested the arterial oxygen saturations so determined, and that these determinations were performed by technicians trained by me," Blalock wrote to Helen in Cotuit.[8]

He was so angry he threatened to quit working with Helen. He was unwilling to continue, he said, when there was "constant fear on your part that you are not receiving due credit." Nor would he stand for her treating him like he was "simply the fellow who is carrying out your ideas."

Blalock laid out additional conditions for an ongoing relationship: Helen would see patients and determine if they were candidates for operation. If so, he would value her view on the cardiac reserve or capacity, but the rest— the time and nature of the operation—was up to him.

Helen opened this letter the last week of September 1945, as she prepared to return to Baltimore for what she expected would be the most rewarding years of her life. She had finally found someone to help her with a solution for her patients, but now he was asking her, in effect, to give him all the credit. Blalock had pointed to the two sentences remaining of the many where she had tried to insert her initials in their original paper as proof that she had gotten credit. These sentences said Helen noticed children died when their fetal duct closed, and her observation, coupled with Blalock's previous animal experiments to move large arteries, had led to the operation. They did not say Helen reasoned that the surgical creation of a blood vessel like the fetal duct would save her patients' lives by delivering more blood to the lungs, or that she asked him to make one. "Where I did deserve credit and had never had it in writing was that the fundamental underlying concept was mine," she told him.[9]

Helen would hardly sacrifice her patients' well-being over a matter of credit. Nor did she believe Blalock was willing to end their relationship if it meant that children would lose out. It was an empty threat, and Helen called him on it. First, though, she sought to calm him. "There are evidently a number of things for us to straighten out" that would be easier to discuss in person, she began her four-page handwritten response. She was sure each of them was "primarily interested in the advancement of knowledge and not only doing our best for these children but contributing to other problems and helping other conditions," she wrote. In her mind, there was no doubt they could go further working together than separately. "You know what is physically possible and I can probably contribute something concerning what is functionally desirable in addition to clinical diagnosis," she wrote. Helen could visualize a surgical solution to help a child with mixed-up parts because she understood how they worked; she believed Blalock could not move ahead easily without her. "I look forward to a good winter & I trust you do, too," she wrote, then reminded him of the funding they expected and the busy month ahead. "Also I look forward with pleasure to seeing you next week." She ended with a nod to Rudyard Kipling's poem, "When Earth's Last Picture Is Painted," likening herself and Blalock to artists seeking truth.[10]

His reaction cast aside everything she had worked on for more than a decade: Although the procedure was novel and Blalock was first to install a bypass of the heart's arteries in a human, it was the product of her work. No one else had observed thousands of patients to deduce what ailed a human heart and, with no knowledge of surgery on blood vessels, suggest an operation to fix it. Her solution evolved from observing how the body naturally compensated and testing it by observing and comparing patients and their hearts. Blalock was trained in physiology, but Helen was fascinated by it. If Blalock did not recall the times Helen had approached him about executing an experimental surgery, or the introductions Helen had made to patients who needed it, perhaps it was because he regularly dismissed ideas that were not his.

Besides calling her "casual remark" about what her patients needed a far cry from making and installing a new blood vessel in a human, Blalock told

Helen that the artificial link between the aorta and pulmonary artery she claimed to suggest to him was not a new idea. Once blue baby operations became public, Gross and Arthur Blakemore at Columbia had told him they had made the same artificial link to increase blood to the lungs.

Other surgeons were experimenting with vascular surgery in the lab but not to reorder the heart. If Gross talked about using his new blood vessel in a patient with a defective heart, it was because Helen had suggested it to him. Worried about embarrassing him, Helen did not mention that she had first tried to persuade Gross to install his duct in her patients until after he publicly recounted her visit. Blalock's comments showed Helen that she would need to document each step in her research to be believed.

*

The diminishment of Helen would upset her fellows—the doctors, many of them women, beginning to arrive on multiyear paid fellowships to study in her children's heart clinic. But Helen's demand for credit and the authoritative way she approached Blalock created sympathy for him among his assistant surgeons—all men. Camps developed.

Blalock expected to get his way. Early in his tenure, he had decisively put down a rebellion by residents over the length of their appointments. He had little experience with women professionals apart from a few classmates in medical school. Blalock flirted with women at parties (wives of residents), put them on pedestals (his mother), and counted on them to open his office window (his secretary). His relationship with Helen would be one of the most challenging of his life.

Six weeks after he threatened to stop working with Helen, Blalock advocated a plan in private talks with Park to reduce her authority in the first of several attempts to gain control over her patients who were candidates for surgery. Blalock was beginning to play a leading role in the growth of Hopkins; he was heavily involved in recruiting department chairmen, including to replace the retiring Park. Hospitals and medical schools were jockeying to hire top doctors and update facilities as laboratory science and surgery exploded. Hopkins was behind in almost every area, even lacking its own bacteriology lab, and using city hospitals to train doctors. The leading candidate to replace Park, Francis F. Schwentker, was an immunologist who wanted to study the link between strep bacteria and rheumatic fever and test the new penicillin. Helen's clinic could become the center of his investigation. Park contemplated integrating Helen's heart clinic with one for adults overseen by the head of internal medicine to give adult doctors more access to children's heart problems. After discussions with Blalock, Park decided he would not get in Helen's way if, as he expected, she was offered a job by a friend at the University of Minnesota. Park even suggested a replacement who, he told Blalock, would "cooperate beautifully" in a reorganization.[11]

If Helen got an offer, she declined it. Was this one of the "many times" in her career she would tell Charlotte that she felt unwanted? We only know that for a few weeks in early 1946, Helen held her breath that the collaboration she had tirelessly sought and finally won might dissolve for a different reason: Blalock was contemplating a move to New York to become chief of surgery at Columbia University. But almost as soon as the rumor hit the newspapers, he announced he would stay in Baltimore following a commitment from Hopkins to build a new seven- or eight-story surgical and research building, a promise of full-time surgeons, and, behind the scenes, a higher salary. "I can do my most effective work here," he told reporters.

Helen lacked leverage to obtain a similar deal. She was desperate for grants to pay the young doctors who wanted to study in her clinic. Park tried hard to find money, but he had little cash flow; pediatric hospitals still lag behind surgery centers. Moving her nationally regarded practice would be difficult while finishing a book, and the one place she might move, Boston, didn't want her. If she knew about Blalock's hefty salary increase, she did not begrudge him. She could live without financial equity. The new surgery building would, of course, benefit her patients. What Helen expected and wanted most was her fair share of credit. And on this matter, she was insistent. She would seize the opportunity to tell her story beginning in the winter of 1945 and into the spring of 1946 when blue babies occasioned a rush of worldwide publicity and media camped out in her office.

*

They gave separate papers at the thoracic surgeons' annual meeting in Detroit in May 1946. Helen had been invited expressly because the host, renowned Detroit chest surgeon Edward J. O'Brien, wanted to underscore the necessity of close collaboration among clinicians, surgeons, and roentgenologist (imaging expert). His own visit to Helen's clinic and Blalock's operating room was the "most amazing experience of my life," he told the group.[12] Whether they shared details of their collaboration is unclear.

Blalock apparently was not present for Helen's talk, and she didn't wait around for his. Afterward, unable to find her, he wrote to say he had heard many compliments about her presentation and that he had tried to call her at what he thought was her hotel.[13] This would be a pattern: avoiding Helen, followed by an apology and a compliment.

A regular part of Helen's relationship with Blalock appeared to be to massage his ego. "Most people are 300 percent convinced of your brilliance," she told Blalock when he passed along a colleague's report that Helen failed to mention him during a speech in California. With humble apologies, she told him her talk followed a half-hour discussion of his surgery by another doctor.[14] Thereafter, she made certain to credit Blalock for developing the operation, often in the first line of her lecture or paper.

Helen's name came first on what would turn out to be their only joint paper after the blue baby operation, on improved oxygen levels in children who received a shunt.[15] Their disentanglement was frustrating to both; Blalock's interests went beyond technical problems, and Helen believed his surgery papers were lacking without her authoritative discussion of the newly diagnosable medical conditions he was fixing. Outsiders refused to decouple them, regardless. After Blalock delivered a solo paper on how and when to install his shunt in an expanding group of oxygen-deprived patients, two top surgeons in the audience, Claude Beck and Elliott Cutler, congratulated "both authors."[16]

Throughout these months, Helen kept their disputes private, or at least closely held, and never complained to staff about her partner. She wanted them to explore the heart the way Lewis and Clark explored the West. Lately they puzzled over a growing number of patients who died on the operating table, including a child from Blalock's hometown of Atlanta. Helen was alarmed. The children's hearts sounded like those of tetralogy of Fallot blue babies; they had the same holes in their heart walls and their skin was blue, indicating they were not getting enough oxygen. But the shunt killed them.

Autopsies showed the position of the major blood vessels in these children, the pulmonary artery taking blood to the lungs and the aorta taking blood to the body, was reversed, a condition called transposition of the great vessels. They did not need more blood to the lungs, but something to stop blood meant for the body from circulating to the lungs in a continuous loop. Until now, Helen had only diagnosed this compromising condition in a small number of newborns, all of whom died. She did not think older children could possibly live with it, so did not suspect it among a growing number of patients seeking her help from around the world. She and Blalock realized the older children had at least one more defect, usually a hole in the heart wall, that ferried oxygenated blood to their bodies and kept them alive. But the variability in the combination of defects made the condition hard to diagnose.

This was an urgent problem. Helen began comparing autopsy results with the moving images of the heart of every child who presented with blue skin and typical tetralogy of Fallot symptoms. Using her fluoroscope, Helen could see if too much blood flowed to the pulmonary artery (right heart) or too little. If blood flowed normally, she began to look for other explanations for their lack of oxygen. The more carefully she looked, the fewer surprises Blalock faced on his operating table. In less than a year, Helen was able to diagnose nine varieties of transposition of the great arteries. She added them to her scrapbook of heart defects.

This was exciting. There were children living with irregularities Helen had long thought fatal at birth. She knew how they survived. Now Blalock could think about fixing them too.

Helen expected to be involved in his experiments early, to explain varieties of the condition she had diagnosed and to evaluate surgical treatments. She was thrilled to hear, in late 1946, that Vivien Thomas and C. Rollins Hanlon, one of Blalock's surgery residents, had begun experimenting in the animal laboratory. Excitedly Helen told Blalock she had lots of patients in her files who might benefit. She also sought to put Blalock's proposed solution for children with transposed arteries on an American Heart Association meeting agenda so she could discuss the many variations and how to diagnose them. "This opens up another field to us—I'd love to talk it over with you," she wrote.[17] Alarmed, he asked her not to discuss his experiments until they were done. She promised to stay mum.

Many years later, one of the doctors who believed Blalock owed his career to the success of Helen's idea and the quality of surgeons he attracted thereafter would call Blalock's effort to disassociate himself from Helen the most notable example of a "petty and ungenerous spirit."[18] Blalock for his part believed God had punished him by forcing him to work with Helen. It was Blalock who preserved their correspondence over credit for the blue baby idea, including Helen's handwritten letters, perhaps so history too would see Helen's demand as unreasonable. In the process, he provided rare written evidence of his own bullying.

By the second anniversary of the operation in November 1946, Blalock had performed 270 blue baby operations, with an 80 percent success rate. Appointments were being made by Helen's office at the rate of two per day and scheduled six months out. Doctors in New York and Boston had begun their own operations, and surgeons from Paris, Montreal, and Shanghai were in Baltimore to learn.

Around the time the pair was first nominated for a Nobel Prize, early in 1947, Blalock was invited to give the main lecture at the American Heart Association's annual meeting, a significant honor. The invitation apparently mentioned Helen, and Blalock sent it to her to see if she would present with him. Helen happily agreed. As was his habit, Blalock suggested she draft a summary of their remarks. Assigning work to subordinates was a way he optimized his time, and his residents would have been honored. Writing was harder for Helen, but she relished the opportunity to define the topic, which would focus on the shunts Blalock was installing in an expanded group of oxygen-deprived children at her suggestion. Assuming Blalock wanted his name first on a joint paper, Helen proposed two papers as they done for the thoracic association. She would speak first, offering an analysis of the heart conditions that could be fixed. He would address the technical aspects of surgery.

Blalock's reply, an outline for a joint paper, infuriated her. She realized he was using her analyses of additional malformations that could be helped by his shunt in a paper under his name first. She would agree to a joint paper only if her name was first. ("I certainly feel my contribution is my original work and I certainly wish my name to come first.")[19] In her ire, she also asked him to change "our" patients to "these" patients. The next day, in a move she said she had long contemplated, Helen asked him to start paying for telephone calls her office made to schedule patients for his surgery on short notice. She described it as a fairness issue.[20]

Blalock waited nine days to respond. Still, he could barely disguise his anger. He told Helen her "amazing" letter contained "certain comments which I think should be omitted." He would put her name first, Blalock said, but he warned her against inserting her initials everywhere she thought was her original work. Then, for the second time in less than two years, Blalock threatened to quit working with Helen "if this friction continues."[21]

Since their old disagreement had resurfaced, Helen replied, it would be better to write their papers separately.[22] He could write about surgery. She would confine herself to diagnostic and clinical aspects. This generated a new battle. Blalock by now saw how difficult it was to restrict himself to surgery-only papers and refused to cede clinical topics or outcome analysis. He suggested Helen limit her papers to preoperative diagnoses or complications, but Helen would hardly entertain limits. From the start, she had linked her analysis of surgical outcomes to techniques and methods. She declared herself free to publish any data arising from her clinic and the hospital and invited Blalock to add his name to any of her papers.[23]

For Helen, this was a blowup to clear the air; wearily she called the matters they discussed in ten letters over three weeks "much ado about nothing." An astonished Blalock could not resist reminding Helen that she started the controversy.[24]

In the end, Helen delivered the Brown Memorial Lecture on June 7, 1947, at the heart association's meeting in Atlantic City.[25] The twelve-page paper under her byline, which she later called one of her most important papers, included analysis of forty-seven operations to install shunts in children with defects like those in patients with tetralogy of Fallot (oddly positioned arteries, for instance). In it she detailed the conditions for success, including the structure of the heart and pathways for blood that would accommodate the altered circulation.

Blalock followed her with a lecture on surgery for these conditions, his presentation marred only by a malfunctioning projector. At a meeting of the American Medical Association the same week, they staged a joint exhibit of their work. By the end of June, Blalock was sending Helen conciliatory notes. His need for patients stopped him from treating Helen as he treated others who challenged him, by banishing them from his inner sanctum. She would try to pick her battles. The phone calls she asked him to pay amounted to $10 that month—about $100 today—and within days Helen softened her demand, promising to limit the calls.

So they managed their frosty relationship to international stardom.

*

On September 9, 1947, Helen and Blalock gave successive lectures in the Great Hall of the British Medical Association, electrifying the staid British audience. Doctors rose from their chairs to cheer when Blalock finished and the darkened curtains behind him parted to reveal a golden-haired child saved by his operation. Blalock with his deputy, Henry "Hank" T. Bahnson, had arrived in London earlier, to begin operating on ten children during a month-long exchange of surgeons between Hopkins and Guy's Hospital. Afterward, Blalock and Bahnson, with their wives, flew to Stockholm.

Helen consulted in Oslo and Helsinki. Then it was on to Paris, where she and Blalock received France's highest civilian award, the Chevalier Légion d'honneur. Dozens of French children had already flown to Baltimore for lifesaving operations, thanks to Aline Pithon, a leading pediatrician in Paris and one of the first to travel to Baltimore to observe Helen and Blalock's work. She had identified seventy French and Belgian children who might benefit from the surgery and issued an "S.O.S." to Blalock to operate in Paris, begging him to bring clamps, needles, and silk thread too since none existed in postwar France.[26] Her collaborator, François de Gaudart d'Allaines, France's top surgeon, was not yet prepared to operate. Helen visited French blue babies she had treated in Baltimore and lectured at the Sorbonne while Blalock operated on seven patients she had identified on an earlier visit. Blalock finally insisted d'Allaines try the operation, thereby launching it in France.[27]

Helen returned home alone, aboard the S.S. *de Grasse*, as the US Congress debated a plan to rebuild Europe and strengthen countries against communism. As was the custom, reporters greeted the ship in New York to interview passengers. In elegant suit, tailored white blouse, and beret, Helen stood next to author and pilot Anne Morrow Lindbergh as she appealed to Americans to send packages of food to France. The French children Helen had examined suffered "not from malnutrition, but from starvation," she told the *New York Times*.[28] In the land of the baguette, there was no bread.

13 PATIENTS FLOCK TO BALTIMORE

In the last months of 1945 and early months of 1946, while Helen and Blalock fought over professional credit, blue babies arrived in Baltimore by the dozens. Helen received a hundred telegrams and letters a week inquiring about the operation. Blalock too was bombarded with requests. In response, they carved up the work. Helen devised an application to be submitted along with blood and X-ray tests. She reviewed the applications, and if she granted an appointment, she examined the patient with her fluoroscope and performed her own blood tests and X-rays to determine if the child was a candidate for surgery. Blalock reviewed her recommendations and operated four times a week. Day to day, they were a coherent team. They sought each other's opinions on which professional meeting invitations to accept. He sent her X-rays and other patient records to evaluate. She boarded an airplane to evaluate sick patients recommended by his friends. Together they worked with doctors around the country to follow postoperative patients.

As exhausting as this period was for Helen, the months after the first successful operations were also exhilarating. She began to attract international acclaim for her role in the surgeries. The first five dozen children were quietly celebrated, their recoveries known to the doctors and relatives of patients who could afford a trip to Baltimore. The nation was preoccupied with the aftermath of war in Europe, including mass hunger and the horror of extermination camps. But beginning with the first anniversary of the operation on Eileen Saxon, Americans began to understand the magnitude and impact of Helen's idea and Blalock's surgery.

"Babies Flock to Baltimore," one newspaper proclaimed in November 1945. Reports of blue baby "miracles" began to mesmerize a public drained by news of war crimes and the cleanup of a devastated Europe. An orphan, two-year-old Bonnie Stewart of Orlando, Florida, was first to make national news. Her autumn trip to Baltimore was financed by neighbors after her father was killed battling the Japanese on Iwo Jima. She arrived at Helen's cardiac clinic in the Harriet Lane hospital in the arms of her grandmother.

On December 4, 1945, alerted by an Associated Press wire service report from Seattle that a toddler was making a cross-country flight to Baltimore for a lifesaving operation, a *New York Times* reporter tracked the little girl's airplane and rushed to the tarmac of LaGuardia Airport, where it had stopped to refuel. There he reported on two-year-old Judith Hackman, the daughter of a tavern owner who, along with her mother and her doctor, had flown 2,736 miles across the country to be examined by Helen.[1] The adorable blond, blue-eyed child seemed fine, but her doctor explained that because of Judith's malformed heart, the girl lost consciousness as often as ten times a day, and without help, she would die within six months. The family's flight was also astonishing for its cost, $5,000 in today's dollars.

A plethora of reporters from morning and evening newspapers waited for Judith in Baltimore. A photograph of Helen smiling, one hand extended toward a cranky-looking Judith and the other reaching into an over-sized black bag, swept the country. For four days, reporters camped in Helen's office and the hospital until the operation.

Judith emerged from the operating room in critical condition, denying reporters a happy ending. While they waited to learn the child's fate, Baltimore newspapers held their readers' interest by reporting on local children who had already benefited from the operation. "Her Life Saved," read the caption under a photograph of one little girl, Janice Ruth Krout, age seven, of Baltimore, decorating a Christmas tree at her school four months after her operation. Stories of former blue babies went around the country via the Associated Press.

For the first time, the nation learned of Eileen Saxon, or E.M.S., as Eileen was called in medical journal accounts of the first three operations. Her death that summer had gone unreported, perhaps because it did not fit the emerging narrative, but now the press portrayed her as a trailblazer. The *Sun* told her life story in December 1945, reporting on the fifty-eight hours she once went without a fainting spell, the weight she gained after the

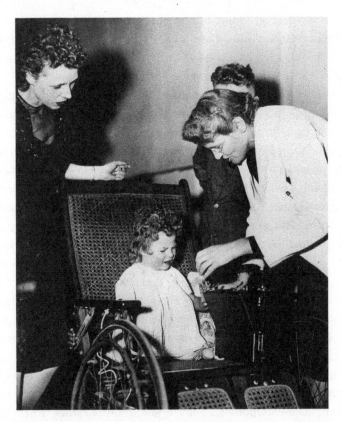

Helen Taussig (right) greets Judith Hackman and her mother, Pauline, after their cross-country flight to Baltimore for surgery, December 1945. Courtesy of the Alan Mason Chesney Medical Archives of the Johns Hopkins Medical Institutions.

operation, the day she stood in her crib, how she learned to walk, her first words, the visits to the park with the Pekinese dog and her father, until the tiny blood vessel Blalock made could not keep up with her body's oxygen needs. Her heartbroken mother told the newspaper that Blalock consoled the family by saying that the operation on Eileen had already saved twenty-nine lives. "Every time I read of them, I think of their mothers and what they must be going through," the *Sun* quoted Mrs. Saxon.[2]

The thirst for news of Judith intensified. Days after her operation, Helen arrived in Atlanta to address the Georgia Pediatric Society, probably as a favor to Blalock, who was being lauded in his hometown, and was besieged by photographers and reporters for the city's two prominent newspapers.

With nothing new on Judith and no blue baby success story in their own city, Atlanta reporters hustled one up in the airport, where a plane carrying a three-year-old girl had touched down for refueling on its way to her home in Topeka, Kansas, after a successful operation in Baltimore.

Then they focused on the woman doctor who had been in the operating room with little Judith Hackman, treating Helen as a hero. Helen used the opportunity to tell them how it all came to be. The result was that while the papers reported on how native son Blalock devised the operation and wielded the knife, it was all Helen's idea. "Hers Was the Theory," a headline in the *Atlanta Constitution* proclaimed. "Woman Originator Explains How Blue Babies Treated." Helen talked of building a passageway to the lungs, the paper told its readers. "I was merely the theorist," Helen said, explaining she had had the opportunity to study blue babies for fifteen years.

There were other women making news—labor reformer Frances Perkins, stateswoman Eleanor Roosevelt, and Pulitzer Prize–winning writers and poets, singers, pilots—but only Helen had made an original discovery in medicine or science. Margaret Shannon, a feature writer for the *Atlanta Journal*, described Helen as a "tall, genial woman who makes explanations of things medical easily understood to a lay mind."

Helen would not have known how fraught her treatment by the Atlanta press might have been for Blalock, who since childhood had never seemed to win his parents' approval. Only a few months earlier, when Blalock learned that the parents of Marvin Mason, whom Blalock treated like his own son, were driving to Atlanta, he asked them to stop at his mother's house to show them Marvin. During the visit to Mrs. Blalock, on March 20, 1945, Lewis Mason spoke reverently about the man who saved his son. To his surprise, she seemed unmoved. Yes, Alfred was great, Mr. Mason recalled hearing, but their other children were just as good.[3] And now the Blalock family was reading in their local newspaper that the surgery that made Alfred famous was Helen's idea.

Nashville was similarly agog a month later when Helen arrived on an emergency visit to examine a sick child. "Originator of Blue Baby Operation Visits Nashville," the papers reported in a headline that could have made Blalock wince. The city where Blalock established his reputation could not get enough of Helen. She was the guest of a prominent attorney friend of the Blalocks who financed her trip to examine a client's child. With press nipping at her heels, she toured Vanderbilt University, Blalock's old stomping grounds, and praised her colleague even as she described her own contribution.

This time when she gave interviews, Helen tried to curb expectations about Judith's recovery, saying that compared with other children, her general health before the operation was poor. Unknown to the media, Judith had suffered a stroke on the operating table. For days she lay motionless while Blalock consulted his friend, the brain surgeon Walter Dandy. Then, unexpectedly, Judith awoke from her coma.

Two months after her cross-country flight, Judith emerged from the hospital in a wheelchair, Helen at her side. Judith had become a symbol of hope at a time when children routinely died from diseases like rheumatic fever and tuberculosis and newspapers tugged at readers' emotions with Dickensian stories of crippled children getting a last visit from Santa. The operation to save blue babies, emerging at the end of a war won with American prowess, marked the beginning of an era when the United States saw itself as morally courageous and technically unstoppable.

At first only the children of wealthy families came to Baltimore, but after Judith's stunning cross-country flight, poor and working-class parents resolved to get help for their children. Louella Champ, a twenty-three-year-old from a sleepy town in southern West Virginia, inspired by the lengths to which Judith's parents had gone to help their daughter, decided she could easily hitchhike 350 miles to Baltimore for her son. On a cold Saturday in January, Champ woke early and dressed her twenty-two-month-old toddler, James, warmly. Like Judith, the boy could not walk. Champ and her sister-in-law took turns carrying the child on their backs as they marched four miles through farmland to the highway. There they waved down a stranger who drove them to Roanoke. A honeymooning New Jersey couple drove them to Washington, DC, then gave them two dollars and bus tickets to Baltimore. In Baltimore, Champ approached a policeman, and in Baltimore's Central Police Station, she and her baby and sister-in-law received warm clothes, food, five dollars, and a bed. The child was admitted to the hospital the next morning, and before Helen could meet them, wire services were transmitting news of their journey.

But it was a community's response to a desperate father that turned Helen into a media superstar. Peter Castelluzzo, a laborer in a wartime explosives and plastics plant in Stamford, Connecticut, approached the *Stamford Advocate* after reading its stories about Judith to learn how to obtain the operation for his own daughter. "I would gladly give up my life if it would do Rose any good," he told the paper. With her mother recovering from an operation, five-year-old Rose attended school sporadically and depended on the janitor to carry her. In the first of multiple front-page stories about Rose, the *Advocate* featured a photograph of her clutching a doll in front of a Christmas tree, her long brown curls framing a bright smile, and explained that her doctor could send tests to Hopkins to see if she would qualify for the operation.

When Rose made the cut for an appointment with Helen, the newspaper announced a fund to finance her trip. Donations poured in, including $1,000 from Rose's kindergarten. On January 12, 1946, "Stamford's 'Blue Baby'" made her way to Baltimore in a donor-supplied ambulance, accompanied by her parents and a reporter, watched by legions of newspaper readers.

Over the next three days, the *Advocate* followed Rose and her parents through the Harriet Lane hospital and provided a blow-by-blow account to readers anxious to learn Helen's decision on whether Rose was a candidate for surgery. In an era before hospitals fielded sophisticated public relations teams to manage the press, the *Advocate* reporter, Bryan Marvin, roamed freely; expertly he extracted details from Helen's staff, including an eight-inch-tall stack of X-rays waiting for Helen's review. In her mail, he noticed applications with postmarks from Montana, South Dakota, Minnesota, and Vermont, and he also interviewed other patients waiting for Helen, including Louella Champ, setting a high bar for fellow journalists in pursuit of the biggest medical story of the era.

While he waited for Helen's return from her emergency trip to Nashville, Marvin introduced readers to the tool she would use to confirm her patients' puzzling conditions. Inside the tiny pitch-black fluoroscopy room, X-ray tubes cracked and snapped around Rose, standing on a narrow platform, while an intermittent picture of her chest cavity was thrown on a screen held by Helen's deputy, Whittemore. Her heart, stomach, and lungs were visible, as was her every "labored intake of breath." The dark room and threatening sounds unnerved Marvin as well as the child's parents, but "little Rose did not seem to mind at all. She turned to the right and left as asked by Dr. Whittemore and her heart was examined by all sides."

Finally, fresh from her flight to Nashville, Helen rushed into her clinic and "immediately took charge" of Rose's clinical examination. "Tot's Fate in Her Hands," bellowed the headline. Rose faces the camera in the photo published by the *Advocate*, locked in a smile with Helen, the little girl's naked chest and skeleton-like arms only partially covered by her long brown curls. Helen ordered more blood tests. "She is to decide today," shouted the next day's headline.

Cooperating with the *Advocate* reporter was purposeful. From her deputy's initial exam, Helen knew that Rose might *not* be a candidate for surgery. Given the rapt attention and goodwill in the Stamford community, she may have accommodated the reporter and photographer to give Rose's supporters something for their money. Helen also wanted to educate the public.

Before the first successful operation, Helen examined six to eight blue babies a year (and witnessed the autopsies of many others). In the seven months after the operation became public, she examined two hundred blue babies. Children with unrelated conditions also showed up at her clinic. The crush of patients illustrated the widespread hope generated by the operation. Parents began to think Hopkins could cure any unusual condition, and the arrival of families without appointments had reached a crisis, stretching the resources of social workers and hospital staff. Helen felt compelled to examine every child, adding to her workload and delaying operations for those who could benefit.

The very morning of Rose's exam, Helen diagnosed James Champ with a spinal malformation and an inoperable heart ailment. Mrs. Champ had received a letter from Blalock urging her to seek medical tests before her hitchhiking journey, but since her husband had lost his job with the railroad, they could not afford them. She came anyway, she told the *Advocate*. "We were penniless, but I wanted to give my baby a chance at life."

The promise of postwar medicine highlighted a problem lingering since the Depression: how people would pay for it. Health insurance was in its infancy, with doctor coverage available to the wealthy or military veterans. Helen shared Champ's view that poor children should have the same opportunity as wealthy ones for the lifesaving operation. To pay for blue baby operations, Helen and Blalock were able to draw from a special fund set up in 1936 in memory of railroad heir Robert Garrett, the brother of Mary Elizabeth Garrett. The Garrett Fund would provide more than $400,000 for children's surgery over several decades. But surgical costs were not the only financial considerations. Families also needed travel funds and lodging for as long as six weeks during their child's recovery, and housing was scarce in Baltimore, which was still overflowing with workers who had moved for wartime jobs. As communities considered raising this money, Helen reinforced her prescreening requirements. At her behest, the hospital issued a public statement advising parents against travel to Baltimore without first consulting a local doctor.

Even then, there was no guarantee, as the public was about to find out. Helen's examination of Rose confirmed her deputy's findings: she was *not* a candidate for surgery. Blood flow to her lungs was plentiful. Helen tempered her delivery of this heartbreaking news to a community rooting for a miracle: there was always hope. She was already studying Rose's condition, she said, and with her exam, Rose had aided the research. The newspaper promised a detailed accounting of the $2,384 in donations; the remainder would be saved for Rose's return visit.

The *Advocate*'s stories created a sensation. Now every newspaper wanted its own blue baby. Michigan City's blue baby came to Baltimore, then Milwaukee's blue baby. Communities in Texas, Florida, and Georgia began raising money for families with blue babies. The family of a Houston five-year-old gasping for breath descended into Baltimore in a B-17 bomber from Barksdale Field, Louisiana, accompanied by an Army nurse. Local companies offered to pay for flights and even limousines from the airport. There were so many children to follow that on January 21, 1946, Baltimore newspapers published a status report, listing the names, towns, and conditions of a dozen blue babies.

*

Whether through her sense of duty, or her willingness to involve patients, or her own canniness, Helen early on became the public face of the blue baby operation. Tirelessly she described the operation to each new set of reporters for newspapers and magazines. She was the ultimate explainer, and in the age of print, she had characters, life-and-death dramas, and children willing to bring their stories alive. Reporters in turn described Helen's patients as smiling, talkative, and sweet. They searched for poignant details such as a father's Army service or the serendipitous way a child arrived in Baltimore. One boy from war-ravaged France arrived after a neighbor read about the operation in the newspaper wrapped around his food relief package.

Children who survived were photographed riding their bicycles, sliding down banisters, and climbing aboard trains. In Buffalo, reporters discovered the second-ever blue baby in their midst and turned her into a front-page story. More than a year after her operation, Barbara Rosenthal, a large bow in her long hair, was pictured next to her bicycle and her younger brother. In the *Evening News*'s account of her difficult childhood and life-changing operation, her mother revealed that Helen had just signed Barbara's permission slip for sleep-away camp.

Stories about horrific deaths assisted by scientific advances gave way to stories that showed how they saved lives. In Baltimore, where evening newspapers catering to the working classes had transfixed readers with blue baby dramas, the cerebral morning paper, the *Baltimore Sun*, assigned a top young writer, Price Day, to the blue baby story during a break in his reporting on Nazis charged with war crimes in Nuremburg, Germany. His February 3, 1946, Sunday magazine cover story on the operation and its consequences was the first of dozens of lengthy articles in mass circulation national publications like *Time* and *Collier's* beginning in spring 1946.[4]

A portrait of Helen by the noted photographer A. Aubrey Bodine on the cover of the magazine evoked one of the underlying attractions of the blue baby story: the postwar prospects for women with professional ambitions who sought to remain in or enter the workforce. In the photo, Helen, her left side to camera, holds an X-ray of a child's chest to the light. She inspects the film, looking for clues; with ever so slight a smile showing off high cheekbones, as if she has solved the puzzle. Wireless framed glasses and finger-waved hair surround her serenely confident face. Her crisply starched white coat identifies her as a doctor. The lapels of that coat barely cover her V-neck dress, accenting an elegant neckline ringed with a strand of pearls. The nails of her long fingers are polished clear; on her left pinkie she wears a gold signet ring. Helen, forty-seven years old, appears graceful and smart, a classic mix of femininity and authority. That she took time from an overwhelming caseload of new patients to sit for this portrait and give multiple interviews in a two-month period showed that she felt an obligation to the public and that she enjoyed what amounted to a victory lap.

Popular press accounts did not focus solely on Helen; they always included Blalock, the "brilliant" surgeon behind the operation, an adjective that Helen often used in her description of his work. The *Sun* magazine story with Helen on the cover included black-and-white sketches of Blalock and Helen inside. The pair was photographed together in another major story featuring three New York children saved by the Blalock-Taussig operation that appeared in the *New York Herald Tribune*. The paper introduced them as a Georgia man and a woman from Massachusetts and described each doctor's role, reporting that it was Helen's idea that blue babies might be helped by surgery and that Blalock had tested the procedure in dogs for another purpose.[5] During one interview, a grateful mother rushed into the room to kiss Helen and then begged Blalock to tell her what she could do to repay him. *Cleveland Plain Dealer* science writer Robert D. Potter reported detailed accounts of Blalock's work but gave the pair equal billing and delved into their careers and the events leading to their meeting.[6] "A woman physician's courageous research and imagination, and the skill of one of the world's great surgeons have combined to bring hope," he wrote.

But it was easier for reporters to poke around Helen's office than Blalock's operating room, and Helen was a novelty. All through the war, women had taken jobs traditionally held by men, and many expected to remain in the workplace and win more opportunities. There were also women readers, buyers of merchandise, and female reporters stuck in women's pages. Helen was their ticket to the front page. Her New England accent, the father who was a professor at Harvard, her mother's interest in botany, and the fact that she lived as an academic: such details revealed a woman with advantages and independent means. One account captured the reporter's excitement that the doctor behind the famous blue baby surgery looked as if she should be running a manor house. Doctors might also be glamorous women!

Women physicians, though still oddities, were not quite the third species described by Osler, the founding Hopkins doctor who had commented famously that humankind was divided into three groups: men, women, and women doctors. Helen's approachability made what she had accomplished look easy and doable and endeared her to the media. Reporters sought her out. Girls from elementary school through college wrote for advice. How could they become doctors? Could they raise a family too? "Many women have combined marriage and a career and have been successful in both—one needs only to consult the annals of renowned individuals and I feel confident that he will find an equal number of married women who became known of [sic] their great achievements or merits as single women," she wrote to one young girl in 1950.[7]

As the teller of the story, Helen actively shaped the narrative to ensure her role was clear. Early news articles all mention Helen's years of examining patients to arrive at the idea for the operation Blalock performed. She touted herself as the theorist, making clear that while she had been thinking and working toward her idea, she did not execute it. "Dr. Blalock thinks I can diagnose any condition," she would say, laughing, "and I think he can perform any operation."[8]

Observing Helen during this time, Vivien Thomas noticed that she seemed extremely happy. He was struck by the change in her from the day in his lab when she had described her patients' conditions. "She no longer had to stand by helplessly and acted as though a great burden had been lifted from her shoulders," he wrote.[9] Helen had always had a purpose, but now she possessed the power to save lives. Along with her uplifted spirit, Helen dressed stylishly, her couture dresses, hunter-green suits, and hats with veils described in some newspapers in the same detail as her patients' ailments. She had begun wearing bright red lipstick in the 1940s. (Elizabeth Arden marketed a Victory Red color to women as their right to be feminine and lovely even while at work.)[10] Frequently, a golden bird or jeweled flower pinned to her lapel or hat highlighted her face. Helen delivered the guest lecture marking the University of Buffalo's anniversary in a velvet evening jacket with rhinestone heart-shaped clasps. Alongside her, a local celebrity, former blue baby Barbara Rosenthal, now a preteen, wore velvet and satin.

Her work was made easier by a decision from the US Children's Bureau to allow rheumatic fever clinics to expand to children with congenital heart conditions. Impoverished parents like Louella Champ could now have their children tested and monitored locally. Helen was gratified; her recommendation that children with congenital heart disease be considered chronically ill helped fund their care. She even enjoyed a hiatus from antivivisectionists who removed their posters from the Hopkins campus, though they would return. She was now a household name. She made it into *Collier's*, one of the era's most popular mass circulation journals. A long article on Helen and Blalock appeared in the same edition as stories on Harry S. Truman's challenging year, famine in India, and the rise of ham radio.[11]

*

Within the Hopkins surgery department, the press's outsized attention to Helen intensified a feeling that she was taking too much credit for the surgery Blalock performed almost daily. Some felt sorry for Blalock. At the all-male Pithotomy Club, an eating club known for its raucous comedic skits about professors, students made fun of Blalock for sharing credit with a woman. Outwardly, he appeared not to mind Helen handling media; he habitually forwarded her newspaper clips, sometimes sent by his mother.

The incessant publicity upset surgeons outside Hopkins for a different reason: it reminded them of quacks' self-promotion. The tabloids' dramatic treatment of Blalock's patients and parents' descriptions of miracles no doubt contributed to that view. Professional jealousy also might have been a factor. To his surprise, Blalock was dressed down in a veiled reference by the president of the American Surgical Association at its April 1946 meeting. William Darrach of Presbyterian Hospital, New York, complained of abundant and unseemly press attention to certain members. Blalock responded forcefully, convening other surgeons to explain the publicity, threatening to resign, and, in the end, forcing his colleague to publicly apologize.[12]

When he returned to Baltimore, though, Blalock sensed an opportunity. He told Helen they should stop talking to the press. They had not encouraged it, he wrote, but "we have simply been the victims of circumstances."[13] Helen agreed. From a practical standpoint, there were more requests for interviews than they could handle. At Blalock's request, she called a reporter and thought she had succeeded in convincing him to quash a story in the *Saturday Evening Post*. "That ought to save us quite a headache," she told Blalock.[14] Blalock himself wrote to a patient's father to try to stop a story in the works by *Life* magazine. But when it came to quashing media stories, these doctors were amateurs. The father Blalock had hoped to sway, a former Army communications officer, was not about to disappoint a reporter friend. The *Life* story ran that summer. The *Saturday Evening Post* story ran in October 1946. If Helen turned down an interview, it hardly mattered; she had secured a place in the public imagination.

*

Helen was promoted to associate professor of pediatrics in July 1946, which carried a small raise. She could not hope to become a full professor—a rank limited to department chairmen, Blalock in surgery and Park in pediatrics. Her promotion brought no improvement in her office quarters, which were so cramped they shocked Genevieve Maurin, the daughter of a prominent bacteriologist accustomed to the grandeur afforded the medical profession in her native France. "The most extraordinary scientific achievements have happened here, in the humblest décor," she wrote.[15] The place where "the idea for a miracle had been born" and more "miracles" ordered daily was a small, nondescript room with a table of books, a floral-patterned covered sofa, and a fiberboard heart.

Fame had brought Helen leverage. Blalock had used his leverage to improve his position. He was building an empire, one that would train doctors who would develop two dozen of the nation's top surgery programs. Helen would use hers to get what she wanted for her patients. She was building a culture.

14 HOVERING

Helen or her fellow called Blalock so frequently with questions or suggestions that he appointed a young surgeon named Denton A. Cooley as the first of a series of go-betweens. Cooley's description of Helen as "hovering" like a mother hen over patients dates to this time. She was on the hospital wards, in the operating room, in the recovery room. Helen also gathered information at autopsies. Once, at least, she even poked into a new laboratory.

In the early months of blue baby operations, both doctors strove to improve patient outcomes, but Helen did not assume Blalock would see the patterns she saw. After Blalock began using the subclavian rising directly off the aortic arch to make his shunt, Helen noticed that it tended to kink up. She concluded it was not safe for most patients and suggested that Blalock use the other subclavian artery—the one springing from the innominate vessel he had used for Barbara Rosenthal—because it was more pliable and easily fell into place. He readily agreed. In the future, he assured her, when the safer artery was accessible and large enough for the patient, he would use it.[1]

This change in the way Blalock operated, using the shorter subclavian artery and sewing its end to the side of the pulmonary artery, would turn out to be the major reason for a dramatic drop in patient mortality.[2] In the first three years of the operation, 1945 to 1948, deaths of children operated on for tetralogy of Fallot dropped by nearly two-thirds. At first, 23 percent of such patients died. Now, annual deaths for all three types of shunts used in operations to bring blue baby patients more oxygen dropped to 6.2

percent. The total number of patients who died on Blalock's operating table was higher, however. Some were older children who turned out to have conditions like transposed major arteries that could not be helped by the shunt. As Helen identified and weeded out such patients, both worked to improve success rates of operations on those who could benefit.

From the first days of the blue baby operations, Blalock told Helen they needed an easier and more accurate way to measure blood pressure and oxygen saturation levels in children with tetralogy of Fallot than what Helen could provide with her hands or Vivien Thomas with his multiple blood tests on the eve of each surgery. This information would tell him something about the capacity of blood vessels and help him decide which one to use to make his shunt and therefore which side of the chest to enter. A more efficient tool also would make it easy for Helen to compare pre- and postoperative oxygen levels. Soon after operations had begun, in spring 1945, Helen lobbied a former classmate, Louis B. Flexner, a biochemist at the Carnegie Lab in New York, for a spare photoelectric cell, a tool Flexner's lab was using on soldiers in Army bombers to measure in-flight blood oxygen levels.[3]

Blalock already knew of something with more potential to assess cardiac output, or capacity to bring blood to the body. It worked by inserting a tiny tube called a catheter into the patient's veins and on to the right heart chambers to extract blood samples and measure blood pressure and oxygen saturation, which doctors could use to more precisely gauge blood flow. From blood flow they could determine pulmonary pressure and how much blood the right ventricle pumped per minute. This procedure, called catheterization, had just been developed at Columbia University's College of Physicians and Surgeons by physiologists Dickinson Richards and André Cournand.[4] It would win them the Nobel Prize. Blalock invited a young physician named Richard J. Bing, an expert in the body's chemical and physical processes who had been drafted as a soldier and assigned to the chemical weapons depot in Aberdeen, Maryland, to set up a catheterization lab at Hopkins—only the third catheterization lab in the country and the first to use catheterization to improve patient care. Bing realized the enormity of Helen and Blalock's work when, crossing the Chesapeake Bay into Baltimore on a ferry in late 1945, he was surrounded by a half-dozen children with blue-tinged skin, all headed to the same place.[5]

Under Blalock's direction, Bing began systematically sampling blood from Helen's patients to learn more about their heart. This patient-specific information supplemented Helen's diagnosis and clarified Blalock's choices in advance of opening the chest, improving survival rates. Once it became routine to measure blood pressure and oxygen saturation for surgery patients, Bing began to show how the tool could be used to diagnose or clarify diagnoses of a wide variety of heart conditions. When Helen sent him a tetralogy of Fallot patient, he recorded the data for the patient and compared it to data from other patients she diagnosed with the same condition. In this way, he produced a set of typical blood and oxygen numbers for each malformation.

This mapping of the oxygen levels and blood flow to a particular malformation was stunning. It provided quantifiable data to accompany Helen's clinical observations. By establishing baseline blood pressure, flow, and oxygen level for each malformation, Bing could, in theory, diagnose a patient by running tests in his lab. In the next six years, he published paper after paper on blood measures doctors could expect to find in patients with twenty different malformations, helping to standardize the study of congenital heart defects.

*

But if Helen hoped to find a supportive colleague in the young physician-scientist, she was sorely disappointed. Bing turned out to be a difficult man who questioned her intellectual capacity when she expressed skepticism about his tool. Brainy and charismatic, a prolific writer and a composer of music, Bing was a favorite of some of Helen's fellows and they studied in his lab with her encouragement. Historical accounts of Bing tactfully avoid describing his temperament, but Helen and her deputies knew it well. The same man who could charm doctors at a Saturday night party by playing his violin could, on Monday morning, move them to tears by angrily throwing his shoe across his laboratory. Rather than personality differences, though, their inability to get along hinged on the different cultures in which they worked, with each being certain that his or her own approach was superior. Bing believed medicine is entirely based on science, and he prided himself in matching Helen's diagnoses to a set of calculations. His boasts that patients could be assessed by a set of numbers rankled Helen, and when she resisted, he turned their dispute into a battle over the question of which of them was the real physician-scientist. He was incredulous when Helen sometimes seemed indifferent to his observations about blood flow and pressure. Her questions led to almost daily arguments in 1946 and 1947, according to a research fellow, Wilfred G. Bigelow.[6] Bing told others that Helen was jealous. He complained to Blalock that she was either unable to understand his calculations or unwilling to appreciate what he was doing. In published accounts, Bing blamed Helen's recalcitrance on her poor hearing.[7]

When Blalock defended Bing and asked Helen to stop criticizing his work, she told Blalock that, in fact, Bing's work had given her a "greater understanding of congenital heart malformations than almost anything else has."[8] But she refused to be intimidated by new technology or bench scientists who believed their tests were superior to her work with patients. What Bing regarded as Helen's aversion to his numbers-driven assessment of cardiac output was skepticism about the catheter as a diagnostic tool and a belief in her own noninvasive methods to diagnose and safely care for her patients.

Throughout 1946 and into the spring of 1947, while she sparred with Blalock, diagnosed additional conditions he might fix, and finished her book, Helen found herself balancing respect for Bing's work with concern for her patients. She knew Bing's information helped Blalock in the operating room, improving the odds for her patients. Once surgery was recommended, catheterization was a critical tool. It could provide precise information on the structure of the heart before the surgeon cut open the chest. But could catheterization alone provide a correct diagnosis? When should it be used? How accurate was it compared with what Helen gleaned from other tools? What was the risk to patients?

From the beginning, the invasive nature of catheterization worried Helen. She maintained a conservative attitude about intrusion into the human body, especially for infants. Tools at the time were rudimentary. She knew about an adult patient in an adjoining lab who had died when doctors had injected dye into his veins. How would a child with smaller veins react to a tube? Helen also feared that a baby or small child would not be able to lie still while a tube pushed its way through her veins and would have to be sedated, an additional risk.

Bing had no experience with patients when he guided a catheter into a child's vein for the first time and asked Vivien Thomas to cut open the vein for him. Both men broke into a cold sweat when the tube they inserted into the boy's heart unexpectedly wound around into his lung. They feared they had poked a hole in the boy's heart.[9] It turned out that the boy had a hole in his heart wall, a condition Helen could diagnose with her hands.

The first 150 operations on blue babies had been performed without benefit of the catheter, showing Helen it was not always a prerequisite for diagnostic purposes. By 1946, Helen had a good idea of its usefulness and limits and began to speak about and publish her findings. In her June 1947 Brown Memorial lecture to the American Heart Association, she differentiated between conditions that could be clarified by a catheterization procedure and those for which it added nothing to what she gleaned from noninvasive techniques.[10] For example, the procedure could be a big help finding an obstructed pulmonary artery to the lung (pulmonary stenosis), but only if the artery could be reached. That was impossible in children with a blocked valve preventing blood from flowing from the heart to the lungs (atresia). In that case, doctors should stick to what they learned from the fluoroscope and physical exam. Catheterization might determine an overriding aorta in doubtful cases. But if uncertain about whether a blue baby suffered from tetralogy of Fallot or the rare Eisenmenger's malformation, Helen heralded a simple breathing exercise Bing developed to measure oxygen consumption. Eisenmenger's patients, who have a normal pulmonary artery, showed an increase in oxygen consumption while exercising. Tetralogy patients, with their compromised route to the lungs, registered a drop in oxygen consumption.

Finally, measurements of blood pressure and oxygen obtained by catheterization could not detect a malformation that even Helen found difficult to diagnose: transposition of the great vessels. For this second largest group of cyanotic heart defects, doctors had to use the fluoroscope; they could recognize it from shadows cast by the unusual position of the arteries.[11]

Ultimately, Helen concluded that the new tool should not be used routinely to diagnose children, and should be used only to clarify mysterious cases. It put undue stress on the patient and carried the risk of a torn vein, prohibiting the procedure if a child ever did need an operation. And since catheterization for routine diagnoses was unnecessary, Helen argued that it was duplicative and therefore wasteful.[12] Hers was an extremely conservative view, and as young doctors anxious to treat adult hearts gravitated toward catheterization and as techniques improved, Helen would be viewed as being late to accept the tool. She never sent infants for catheterization and stopped sending Bing young children for blood tests once she determined they were *not* candidates for operations. Park too believed the tool could only supplement, not displace, Helen's methods.[13]

But specific puzzles prompted Helen to seek Bing's help. When a five-year-old girl appeared in Helen's clinic in late 1947 with a small aorta and a large pulmonary artery that reached over the wall separating the bottom chambers of the heart, Helen ruled out transposed great vessels because she saw from her physical exam that the pulmonary artery arose as usual from the right ventricle. In addition, the child was older and in better health than most other children with transposed arteries. While lecturing in Norway a few months earlier, she had heard of a girl with the same unusual features. In that case, Helen suspected but ruled out another malformation, Eisenmenger's complex, because the patient had been blue from birth. Helen decided her uncertainty justified catheterization.

When the first test in Bing's lab didn't clearly indicate a diagnosis, Bing sent the child for a second test, an angiogram, in which doctors in another lab used a catheter to observe dye flow along with blood into the veins. A machine in the lab was not working that day, and doctors injected dye into her blood three times within forty-five minutes to try to obtain a good film. The rapid injection of dye into the circulation system altered resistance and reduced the amount of blood flowing to the pulmonary artery, shutting off what little oxygen had been able to get through to the body. A few minutes after the last injection, the girl suddenly lurched upright and fell back down, dead.

An autopsy revealed that this child's aorta was on the opposite side of the heart, so that both main vessels emerged from the right chamber and only a small amount of oxygenated blood could have reached the aorta from the left side of the heart. This combination, a normal pulmonary artery but a misplaced aorta, was a malformation Helen had not seen, a variation of transposition of the great vessels. It is known today as the Taussig-Bing anomaly.

In discovering this variation, the doctors showed that catheterization could not always provide a definitive diagnosis, and invasive journeys into blood vessels sometimes could be fatal.[14] Not long after, Helen lost a second patient during a catheterization procedure using dye. This time, the only route for oxygenated blood flow to the child's lungs was an extremely narrow vessel, and the dye pushed out the entire blood supply through that passageway, cutting off oxygen. In a joint paper describing the rare malformation, Helen and Bing warned against injecting dye into patients with low oxygen or obstructed vessels and no alternate source of oxygen. The experiences reinforced Helen's cautious nature.

Helen wrote two papers with Bing despite their tortured personal relationship. The screaming coming from behind her closed door one day in 1950 or 1951 led an alarmed staffer to worry, only partly facetiously, "They are going to kill each other." A prolonged silence followed, during which the doctor wondered if they had succeeded. This time, Bing was upset that Helen had mentioned Blalock in a published paper but not him.[15] Bing often complained that his contributions were overlooked; he expected his name to be linked to those of the famous doctors.[16]

Helen would have been horrified by this outburst fiery enough to be heard outside her door. Perhaps during the quiet interlude, she promised to include Bing's name in future papers when she discussed diagnostic tools; his name had appeared in previous papers and would later appear in her book. Perhaps she praised his work the way she praised Blalock when he was angry. Whatever she said worked; when the door finally opened, out they came, smiling, arm in arm.

Their relationship remained contentious, however, because Bing repeatedly told Helen she was wrong about the cause of the problem facing blue babies. He argued that some blue babies suffered not from insufficient blood flow to the lungs but from insufficient oxygen in the blood—a problem of quality, not quantity. The concept of effective blood flow to the lungs that Bing established was significant; it prompted doctors to assess oxygen levels in the route to the lungs as well as the heart; if oxygen was high, as it would be in children with transposed arteries whose blood constantly recirculated to the lungs, increasing blood flow would not help.

Helen acknowledged the concept even as she resisted catheterization of the right heart, which pumped blood to the lungs. But there was no denying that blood had to reach the lungs to pick up oxygen, and Helen's description of insufficient blood flow and her solution worked for children with narrow or blocked routes to the lungs (pulmonary stenosis or atresia). She could tell who benefited by looking at the size of their arteries with the fluoroscope.

*

Helen's philosophy became the prevailing attitude among pediatric cardiologists who nonetheless embraced catheterization as an important additional tool. Pediatricians abroad and at leading US institutions also resisted the invasive technique, both because of potential complications and fear that it would diminish the importance of the physical examination, which already allowed them to measure pressure in the pulmonary artery.[17] By 1952 pediatricians' refusal to refer fragile babies for catheterization was widespread, not least because of Helen's analysis in medical journals, but also cautionary notes from such figures as Charles T. Dotter, the first doctor to invent ways to use the catheter to treat patients rather than simply test their blood flow.[18] Even Bing acknowledged in a 1952 paper the primacy of the clinical examination. In his lab, he was kindly toward children and put them and their parents at ease with his banter, even engaging his assistant to sing to the older ones to avoid having to sedate them. (Decades later, when Bing wrote to tell Helen his new job allowed him more time to compose, she told him wryly she "always wished you had spent more time with your music."[19]) He left Hopkins in 1951, after six years. Blalock tired of him and Bing, bored by routine work, was beginning pioneering work on cardiac metabolism, or energy sources of the heart.

*

Mary Ellen Avery, who graduated from medical school in 1952 and would become the first woman to head a clinical department at Harvard Medical School, recalled the excitement of this postwar scientific era, when pediatricians began to measure and investigate a variety of problems and "great advances were being made before our eyes." A fan and friend of Helen, she felt "embittered" at first when Helen's skills with the hand and fluoroscope got "pushed aside" as catheterization came into vogue. But she quickly understood that measurements gathered through catheterization were transforming the cardiac world.[20] When surgery advanced in the mid-1950s so that the heart itself could be repaired, these measurements would become vital.

Helen did not feel pushed aside. She anticipated that new technology would vastly improve patient care. She appreciated and would use new tools when she judged them necessary and safe. She viewed Bing's accomplishments the same way she viewed Blalock's blue baby operations: Bing had a tool, and she had shown him children with defects she had identified so that he could measure them. Without her diagnoses, doctors who recorded differences in oxygen and blood pressure would have no idea what they meant, she would say, and it would have taken them years to find out. To Helen, the numbers Bing obtained from the catheter were in no way superior to the discoveries she had made with her hands and the fluoroscope. Her patient-based physical investigation of heart malformations was no less a scientific approach than his diagnosis by catheter in the laboratory. Working in a lab did not make anyone a scientist, in Helen's view. The essential tool of the scientist was not the hand, the scalpel, or the catheter. It was the brain.

Helen made this point in dramatic fashion before a dazzling array of women scientists employed by the country's top academic and government institutions. The occasion was the opening of modern biology and chemistry facilities at the new 287-acre campus of Goucher College. The renowned Maryland women's college had prepared women in chemistry and physics since 1885 and launched doctors like Margaret Handy and Ella Oppenheimer. To advance scientific knowledge, Helen told the largely female audience, they needed more than new facilities or the skill to observe and deduct or run controlled experiments in laboratory. To understand what they are observing—to find truth—they also needed imagination, she said. This was high Emerson. "Science is the repeated process of observation and deduction, of observation and then hunting the inner meaning and afterwards, if you get a clue, test it, to determine whether the idea is correct," she said. "The fundamental inquisitiveness of seeing a thing and trying to see what its inner meaning is, its relation to other things, the idea back of it, and to be able to understand something from what you see," she said, "that is the essence of all true science."[21] For Helen, the question was not who is a scientist, but what problems scientists should try to solve. And who should benefit?

The audience would have been familiar with the name Helen next evoked, her friend Emily Dunning Barringer, the first woman surgeon, who had been mentored by the crusading doctor Mary Putnam Jacobi. "Women go after what is right with no consideration as to whether it is practical or not," Helen said Barringer had told her. "Of course we do—right is right. It is not limited by practicality or feasibility or anything else," Helen said. "If the problem in which you are interested in is right, it is an objective and goal in itself," she said.[22] Eventually, by following scientific methods, they would find the answer. Then she asked them to consider problems *created* by postwar scientific discoveries such as hers—ones she faced daily, such as how doctors decide whether an individual should risk heart surgery, or wait until technology matured, or how to pay for treatments. As Helen spoke, the country was engaged in yet another debate about how to pay for health care, including hospitalization for the elderly. The advancement of science would have no effect if most people could not afford medical care, Helen said.

Five hundred people heard Helen talk on this windy but warm spring day, April 3, 1954. Her "vivacious charm" apparently made up for her scant attention to the new science facilities. To one professor in the audience, it was apparent that "a deep moral conviction forms the foundation of Dr. Taussig's life and work."[23]

*

Duty and methods led Helen to stray into others' realms to watch over patients. But even she could not have predicted Bing's reaction the day when she ventured inside his laboratory. She had been summoned by Charlotte, who after her fellowship with Helen studied in Bing's lab. On this day, while Bing was at lunch, a child developed an irregular heartbeat during a catheter procedure, and it did not revert to normal when Charlotte removed the tube. Unable to reach Bing and fearing the child would die if her heart failed to push out enough oxygenated blood, she telephoned Helen. Helen rushed to the lab, and together they stabilized the baby with drugs. Helen lingered to be sure the patient was comfortable and to calm Charlotte. On her way out, Helen bumped into Bing. He brushed past her and in a scream that echoed down the hallway, he demanded: "What is *she* doing in my lab?"[24]

15 MAP TO THE NEW WORLD

In the introduction Helen penned in the summer of 1947 to her book on the problems of hearts, she called it a reference to help children's doctors diagnose heart malformations. In his foreword, Park described it as so much more: it was a map to a previously unknown land and to the surgical possibilities in this new world.[1] For over a decade Helen had used her hands and the fluoroscope to identify malformations. Here it was: everything she had learned, in 572 pages. The book people told her was a waste of time because it described conditions that inevitably led to death. Arriving when it did, just after the first successful operation to alter the way the heart works, her timing was perfect.

Helen's book explained how the deformed heart works, discussed malformations that caused poor oxygen flow to the lungs, malformations that did not affect oxygen flow, and treatment. She compared each problem to the normal heart and circulation. She provided a description of how and when each of these deformities occurred and how they altered the circulation, and she explained how children's bodies compensated for deformities at birth and as they grew. Malformations were not isolated hit-or-miss events. Each one occurred again and again in the population, during fetal development, changing the size and shape of the heart. These changes were clues. For each deformity, a physician who knew what anomalies to look for in the living patient could diagnose them. Using images of her own patients, she showed readers what these changes looked like under the fluoroscope. For example, the heart assumes a different abnormal shape depending on whether the flaw occurred in the lower right chamber or ventricle, which moves blood to the lungs, an obstruction called pulmonary stenosis, or if the right ventricle is enlarged or missing altogether. When there were multiple variations of a defect, Helen described shapes and processes for each. She described how deformities appeared infant to infant and how the malformation looked in the adult, when the body had had time to develop around the problem. The book's index alone, listing heart defects and their variations, ran thirty-seven pages. It included forty color illustrations and several hundred drawings.

In explaining fetal development, Helen credited research by her cousin, Marjorie Prichard, and Prichard's colleagues at Oxford, whom she had consulted long ago beginning with X-rays strewn over her bedroom floor in Cotuit. She also paid homage to Maude Abbott, the doctor whose fundamental pathology on defective hearts helped Helen diagnose these deformities in actual patients. In her nearly forty-year career at McGill University, Abbott had assembled the largest collection of flawed hearts in the world. She published her own path-breaking book, *Atlas of Congenital Heart Disease*, in 1936. By integrating the diagnosis of deformities in children with laboratory descriptions of misshapen hearts by Abbott and solving some of the mysteries Abbott laid out about how these hearts worked, Helen's book, *Congenital Malformations of the Heart*, became the most important current reference on the heart.

At the time of publication, surgeons could perform only three operative procedures for heart problems. They involved vessels *outside* the heart. The first two involved repairing existing vessels to improve blood flow and thus the heart's performance (closing the fetal duct, accomplished by Gross in Boston, and removing an obstruction from the aorta, done by Clarence Crafoord of Sweden). The third, Blalock's shunt, rerouted the circulation of blood to the heart and lungs and suggested the possibility of many more such body-altering operations, possibly even *inside* the heart. Helen assumed so, anyway. She included ostensibly hopeless heart defects, even one in which the heart protruded through a hole in the breastbone. This condition was easy to diagnose, she wrote, but was merely a "curiosity until surgical skill is able to replace the heart within the chest."

Because she wrote from experience with patients and in simple style, the book is easy to read. Her prose is precise, logical, and engaging and her tone accessible, something reviewers at the time appreciated. It showed her to be a skilled teacher, one wrote, adding, "This book bears the mark of years of patient, careful work and is a tribute to the purposeful specialization of one of the outstanding clinicians of our time."[2] Helen's goal was to help readers become as good as she was at diagnosing these conditions, so they too could save lives. "Accurate diagnosis will become progressively more important as surgical skill is better able to correct or compensate for the various malformations," she explained.

The book did not include all malformations or discuss surgical solutions other than for tetralogy of Fallot. Helen referred readers to papers by Blalock and others, since corrections of these problems were rapidly progressing and operations "may be subject to radical revision." Indeed, one surgeon had already revamped Blalock's procedure.[3] Despite these limitations, the book was a sensation. A year and a half after publication, Helen's book entered its fourth printing. When the type on the printing press wore out, making Helen's book officially out of print, 13,081 copies had been sold—an enormous number for a textbook and astonishing given there were fewer than five hundred pediatricians in the United States. Physicians treating adults' heart problems would want to know about congenital heart diseases in children, but Helen's map would have been of greatest interest to surgeons who aspired to fix them. The bookstore was the first stop for doctors arriving to learn from Helen and Blalock; men in white coats carried her book as they walked the streets near the hospital.

Instead of being outmoded by the catheter, Helen made Bing's work important. Doctors evaluating the book decades later expressed awe that Helen made her diagnoses without any of the tools used today. Like the museum visitor viewing a Roman statue, we wonder how the artist created such art, much less carried the stone from the quarry to carve it. Helen's too was an original work of art and science. Modern technology makes it possible for people to skip steps, so that everyone does not have to be an expert in every detail. She made her discoveries by observing, thinking, and touching. She made them with her hands and hard work.

*

The celebrations over her book came on the heels of the retirement of Helen's mentor, Edwards Park. He was seventy and tying colorful fishing lures in his beloved Margaree in Cape Breton. Park had extended the lives of untold children through his work on rickets and had brought Helen and Blalock together. Helen's sadness was ameliorated only by the prospect of annual visits. He stood in for her father and had lived to see her success; for that, she would always be grateful. No one could tell her what to do except Park, and he had advised her sparingly.

Helen was not in the running to lead the pediatrics department. Instead, she was shaping a new medical specialty: pediatric cardiology. In the previous eighteen months, to cope with an overload of patients Helen worked long into the evening with only two assistants. To run her clinic and supply other medical schools and hospitals with children's heart specialists, Helen needed to train at least five or six doctors annually. Between 1945 and 1947 she had approached the federal Children's Bureau, the Commonwealth Foundation, and elsewhere for research and teaching grants. No longer a repository for women denied other opportunities, pediatrics was booming, thanks to specialized clinics and the chance to apply scientific methods. One of the doctors who had managed the first blue baby operations, Whittemore, was returning to Yale University to open the first congenital heart clinic outside Baltimore. Helen's book prompted doctors from around the world to vie for fellowships, to learn from her to diagnose heart ailments and return home to develop the new academic field she had founded. They in turn would provide Helen the help she needed to handle the onslaught of children seeking diagnoses and operations. She wanted pediatricians training in cardiology—fellows, they were called—to stay two years, and she spent long hours soliciting and selecting candidates through trusted colleagues in the United States and throughout Europe.

*

As she sent final chapters of her book to the publisher in 1947, Helen also acted to secure her personal footing. She was almost fifty years old. She had hoped and always assumed she would return to Boston, but success cemented her to Baltimore in a way that she could not imagine. On weekend afternoons she began to explore the streets of Baltimore in search of a house of her own. A house would provide ballast against the intensity of her devotion to her patients and the difficult aspects of her work, including battles with Blalock and Bing and Schwentker, Park's successor, which affected her spirit and her self-confidence.

"It is . . . the balance one attains, and one's relation to people that is so different when one has a house instead of living in a single room," she would write Charlotte. "Many people do not realize how important it is for the unmarried woman, but truly it is more important for she than for the married woman who inevitably puts out roots and branches."[4] Her idea of balance was not to separate herself from work but to expand her sphere of influence. She believed personal connections were key to professional ones. She had witnessed the camaraderie that developed at her father's dining room table as he established the discipline of economics, and she experienced it at Park's table. Now it was her turn to preside.

Her inheritance played a role in her timing. The median price of a house in Baltimore was $8,033. With an annual salary of $6,000 or less during her most productive years (a recent raise brought her to $10,000), she could not have afforded one on her own. This was also true for male academics at her level. Those who were not independently wealthy but wanted homes and families left academe for private practice, where they earned twice as much. They settled in leafy neighborhoods of English-style stone cottages in Homeland, the shingled large family homes in Roland Park, and stately brick mansions of Guilford favored by the city's elite. Helen was not part of this social set. She wanted something more private, away from extraneous background noises that made it more difficult to distinguish sounds. She wanted land for Spot, a mutt with floppy ears, large brown spots, and lanky limbs, who acted as her scout, alerting her to visitors. Helen had used an amplifying stethoscope since at least 1936. By the late 1940s she also used hearing aids; the first was a box she hung around her neck and aimed at the speaker. Then she clipped a gadget the size of a transistor radio to her dress. This was probably a radio that connected binaural hearing aids and allowed her to hear sound from both ears simultaneously. It was so heavy it pulled on her clothing, and sometimes she was so preoccupied by patients she didn't notice when the tugging exposed her chest, leaving her "almost indecent."[5] The next version would be less cumbersome.

For the house itself, she had in mind something modest, like her cottage in Cape Cod. Simple and full of light, the regional style featured a central chimney affording fireplaces in the living room and first-floor bedroom. She already possessed the architectural plans and would adapt them to local costs and material. The Baltimore version would be smaller, stucco rather than clapboard, and topped with slate shingles instead of the traditional Cape Cod cedar. The only remaining task was to find a suitable setting. It should be serene, allowing her to cultivate plants and immerse herself in beauty, preferably with access to water so she could swim or canoe. It should be in the country, with a bountiful lawn for games and summer barbecues, but near enough to her office downtown. The neighbors must be educated and conversant in the art of ideas, but respectful of her privacy.

One day, several miles east of Park's first Baltimore home, she turned onto a narrow dirt road, Hollins Lane, that sloped upward, and when she got to the end, she stopped the car and got out to look. Here was six acres of land at the top of a bluff overlooking Lake Roland, a man-made reservoir surrounded by hundreds of acres of parkland. The property had been abandoned and would require cleanup, but she was not the first to recognize its potential; a small hotel on the site, once the summer mansion of a railroad baron, had burned to the ground during the Civil War.

Back at her desk in the apartment that had become cramped, with a surveyor's map of the land in her hands, Helen penciled in where she would build her cottage and garage. She positioned her first-floor bedroom in back, facing the lake and the rising sun.[6] Attached to the bedroom would be a screened porch, with a brick floor, where she would sleep. With her poor hearing, Helen could scarcely hear birds and other creatures from the porch or through open windows. But once she nurtured this barren landscape, she would be able to see and smell her surroundings. Every window would eventually possess a delicate view. From the living room, it was a euonymus or burning-bush tree, whose seed-filled pink berries attracted birds. From the dining room, seated at her father's table, between the trees and past the azalea bushes, she could see the lake. This was as close as she could get to her beloved Cotuit.

16 A HOUSE OF HER OWN

Helen's was the largest heart clinic in the world, with appointments obtained months in advance and parents who traveled long distances with perilously ill children. It was besieged by desperate patients but also by physicians and surgeons, including Robert Gross, who, having rejected Helen's idea for surgery long ago, returned to Baltimore to see what he could learn.

The deluge of patients and visiting doctors rotating through her clinic put quality control on center stage. Helen wanted her fellows to succeed, using her methods. But she did not want her patients to receive anything less than the top-quality care that she herself could provide. She and her fellows were applying what they learned from one patient to the next to establish what doctors today call best practices for diagnosis and treatment. As she gathered more information, she began to see surgery in a new light, the light of success, with a giddiness that emboldened doctors and encouraged patients.

Whittemore, her first fellow, described the clinic atmosphere as intense.[1] Fellows were expected to examine and diagnose a patient in one day. "Fellows saw the patients first," recalled Anthony (Tony) Perlman, who traveled to Baltimore from South Africa to learn from Helen after reading her book. "Then at a conference with the patient, [Helen] would be introduced, hear the history, findings, and examine the patient in front of the family." Then the fellow would present the results of his or her examination. Helen listened carefully and, afterward, with staff, she would be straightforward about her own conclusions even when they differed. "I think you are all wrong about that," she would begin.[2]

Frequently, as Helen assessed information arriving from the operating room and hospital she revised protocols for her patients, for example, on when to send patients for catheterization.

Helen's low tolerance for risk also revealed itself in her recheck of each diagnosis, even those by Whittemore, whom Helen trusted to run her clinic. This would continue until the mid-1950s when Helen installed a full-time deputy. By reassuring herself, Helen boosted the confidence parents had in their own difficult choice. From Dorothy Saxon onward, parents forced to make decisions in the absence of information routinely expressed their trust in Helen.

Her relationship with patients began before they arrived, when Helen or her staff arranged housing, translators, and, for a parent traveling alone, a local advocate. Her approach was holistic; she knew from her own child-hood that the patient's attitude was key to his or her healing and sought to understand each family. Some of her patients, like one young boy who was the only surviving male in a Jewish family that perished in death camps, experienced suffering beyond their illness. Parents too were traumatized. Geneviève Maurin brought photographs of her bombed home in Caen, northern France. To keep her son alive, she carried him on her back and foraged in woods near encamped American soldiers.

Helen personally instructed fellows on the importance of greeting and comforting patients and their families on the day they arrived. Her own reception became the model. She gave patients—and parents—what her father had given her: all the time in the world. When she spoke, it was right to them, without artifice. She listened intently. She smiled. "She would quiet them down if they were upset, answer every question, and sit there as long as needed," recalled Norman J. Sissman who after his 1958 fellowship with Helen built the pediatric cardiology department at Stanford University. "That is really how I learned."[3]

Helen's physical connection to patients began with her hand on their chests. On their way out, she hugged every patient and planted kisses on their heads or cheeks. This was not easy for her; Helen was not someone who hugged friends. She could be distant, even cold. To Perlman's wife, Patsy, who was a child when she met Helen, Helen sometimes seemed as if *she* could use a hug. But Helen treated her patients like family. "Come visit me," she told children. And they did.

*

Her fellows knew Helen had become adept at palpation because of her hearing loss, and many watched in awe as she silently placed her hand on a child's chest to feel its movement, holding it there a long time, as if in a trance. This was the pre-ultrasound era, before modern tools could instantly reveal the heart beating, a time when medical students still prided themselves on their ability to feel a spleen or hear a diastolic murmur, often returning to the hospital at night to practice. Helen's sensitivity to vibration was so extraordinary that she could place her hand over the hand of another doctor as he examined a patient and, if their hands were placed correctly, she could feel the vibration *through* his hand.[4] This was unique; even trained doctors could not feel the vibration.

Few doctors could hope to match Helen's ability with her hands. Helen knew this; she encouraged medical students and her fellows to use all available tools and to learn catheterization. But she insisted they examine patients first with their hands. Helen believed it ensured a correct diagnosis. Only somewhat jokingly she told young students that while they might lose their stethoscopes or, as in her case, their hearing, they would never lose their hands.[5] Mainly she wanted her fellows to experience the deep connection she felt with her patients. This human connection also was a check or counterpoint to help doctors evaluate technology.

She taught her method in a way they could not forget. At the start of each rotation through Helen's clinic, when second-year medical students brandished new stethoscopes and were anxious to learn how to use them, she would single out a student and beckon him to the examining table, inviting him to listen to a child's heart. Robert I. Levy nervously approached the table one day after Helen called on him. He had only gently put his new stethoscope on the child's chest when he felt something grab his hand and swipe it away. When he turned around, Helen was holding his prized new tool. "The hand before the ear," she told him sharply.[6]

Equally intimidating was watching Helen use the fluoroscope. This was the moving X-ray, a screen that projected fluorescent light as Helen maneuvered a baby to watch his heart move and pulsate at different angles and gauge the size or location of a ventricle. To waste no time assessing squirming children in the pitch-black fluoroscopy room, Helen donned dark glasses and walked around her clinic to acclimatize herself. It was a detail some found exasperating, but also the reason Helen was usually right. She would observe the working hearts of five patients in succession, narrating what she saw—often what her fellows did not see—diagnose or pose a question to resolve and then return to her office and dictate reports for each child from memory. This process helped her to catalog these images and fix them in her memory. When a child returned for a new exam, she remembered not only the sounds of that child's heart from the previous visit but also how it moved.

*

After learning to diagnose first with the hand, the second unusual requirement Helen made of doctors training under her was that they follow their patients into the operating room. Observing operations helped pediatricians learn about surgeries and provided an opportunity to get to know Blalock and his residents. Part of a fellow's job was to know what could be done for their patient and to ensure that Blalock or his team member "was doing things you thought the best way," recalled Sissman. At subsequent staff conferences in Helen's clinic, fellows reported on what they saw, including errors. "If there was a problem, she wanted to know what we would do, and that we *would* do something, whatever we could to help the patient," Sissman said. Sissman's 1958 fellowship coincided with the residencies of top cardiac surgeons of the era—Hank Bahnson, David C. Sabiston Jr., and Frank C. Spencer. Helen was teaching her fellows, many of them women, not only to make decisions on whether to send children into the operating room but also to hold their own against young surgeons with great talent and egos to match. All the surgical residents were men.

*

The way Helen treated vulnerable patients stood in stark contrast to her brusque impatience in her clinic when a doctor failed to do something in the way she had instructed or expected. On patient rounds, one of her fellows always carried Helen's amplifying stethoscope, a heavy contraption in an 8-inch-by-6-inch wood box. One afternoon when Helen put it to her ear, warmed up the tip, and put it on the child's chest, no sound came out. She turned around and snapped at Sissman, the fellow carrying it that day. He had forgotten to turn it on. These rebukes were jarring because they occurred amid Helen's soft-spoken and gentle treatment of patients. "You saw how beautifully she treated the patients and five minutes later she would chastise a doctor because they wouldn't do something," recalled Perlman.

She also picked favorites, elevating those who excelled and had winning personalities to be her go-to assistant. But being out of favor was no worse than being in favor since those in favor worked even harder to accommodate Helen. "The fellows took care of her, they were solicitous of her, because she demanded it," Perlman said. One stern look and Helen's pursed lips was all it took for fellows to drop to the bottom of the heap. Helen was particularly harsh on male doctors who fell down the ladder. "It was her tone of voice," Charlotte remembered. "Helen would be all over him. She could be nasty." During the six years she worked under Helen in Baltimore, Charlotte never saw her reproach female doctors in this fashion.

Although she selected her fellows, Helen did not always like them. Certain people or behaviors irritated Helen and she could react harshly. Michel Mirowski, a Polish-born refugee even more impatient than Helen to succeed, had asked her permission to share an innovative paper at a gathering of her fellows. Others looked forward to hearing it. Whether because of her expectation for clarity in the workplace, her difficulty lipreading a foreign accent—it was like learning a new language—or because she found him intimidating, Helen told Mirowski he first needed to improve his English. Day and night he worked to please her, with the same determination that would lead him to develop the implantable defibrillator, a machine to reset hearts, prolonging millions of lives. Helen's treatment of Mirowski seemed to be a kind of test; in later years, when half her staff came from other countries—indeed she welcomed doctors from underdeveloped countries— Helen adapted to *them*, by learning short-cuts to lipreading.

After Mirowski improved his English and presented his discoveries, she collaborated with him on three major papers.

*

Helen was insistent on an orderly and smooth operation in her busy clinic and reprimanded staff who did not keep her moving. They went to great lengths to be sure she was not interrupted. One day, a ten-year-old boy and his mother arrived in the clinic without an appointment, saying Helen had invited them. Helen's fellows told the pair that Helen was too busy to see them, but they refused to leave. Helen's staff reminded her they were defending her schedule by not making room for unplanned visitors. But when Helen learned that the boy outside was Marvin Mason, the third blue baby, she ran from her office with arms outstretched. "How lovely to see you!" she told mother and child. As Helen hugged and chatted with her patient, her fellows exchanged anxious glances and scrambled to keep her on schedule.

Helen's infinite patience with families was the reason her clinic routinely extended into the evening, until 7:00 p.m., more than two hours after the official closing time. In summertime, after the last patient had left, Helen's sweltering staff would jump into one of their cars and head over to a Route 40 crab house. Reeling from the intensity and Helen's exacting standards, they would order mounds of crabs and pitchers of beer and unpack their day.

For surgical residents who hoped to specialize in children, Helen's high-pressure clinic with its emphasis on collegiality was a relief from the belittling they could expect during their six-year program under Blalock. His practice of winnowing down the class of a dozen or so interns every year until only one remained to be chief resident was "inexcusably brutal," according to one chief resident, Frank Spencer.[7] (Rejected residents became stars elsewhere, sometimes with Blalock's help.)

On patient rounds, Helen was not as loquacious as Blalock. She did not joke with students or tease them. Nor did she "put us down," as Blalock did, for example, addressing Francis D. Milligan as "Mulligan" and telling the vulnerable young doctor, who was training in gastroenterology, "I know you didn't get that."[8]

Blalock's charisma overwhelmed his critics and softened the blow of his iron hand. Helen had grace; with impeccable manners she disguised her tenaciousness.

The spirit that developed in Helen's clinic and spread to the discipline of pediatric cardiology emerging from it reflected admiration for the high bar she set for herself. Few fellows could bring themselves to hug patients or attend funerals as Helen did. As for her demands and unexpected rebukes, all was forgiven because her motivation was clear: to help patients. This was illustrated in an extraordinary way one day in summer 1955. Helen was working in her garden in Cotuit when her thoughts turned to a boy she had diagnosed eight years earlier with pulmonary stenosis, or narrowing of the artery that takes blood to the lungs. She had not examined him in five years. Based on what she now knew, she realized her diagnosis was wrong. Instead, he had aortic stenosis, reduced blood flow to the body because of a narrow opening at the aortic valve. With this injury, his heart worked harder to push blood to the body. It would harden up with all that work. She ran inside and dialed her office, asking the fellow on duty, Henry A. Kane, a Delaware pediatrician, to find the patient. With help from Helen's secretary, they tracked the boy, now in his late teens. An examination confirmed the new diagnosis. The boy's condition had worsened: some of the blood was now leaking back to the left ventricle instead of moving into the body (aortic regurgitation). He would be among the first patients to benefit from new operations on aortic valve flaws.[9]

*

Helen's warmth and genuine nature were such that doctors who were targets of her occasional rebukes also did not think twice about socializing with her. More than socialize with her, Helen's fellows fell under her spell. In autumn 1949 after she moved into her new house on Hollins Lane, Helen began hosting Sunday dinners in the manner of Park and his wife Agnes. It was a dinner invitation with a twist. She would be working outside on her grounds on Sunday afternoons, her only free time. Would any of her fellows like to come a few hours early, in work clothes, to join her in some yard work? For the next few years, half a dozen young doctors showed up at Helen's house on Sundays with a change of clothes. Helen was extremely hospitable, and those who attended enjoyed themselves immensely, returning Sunday after Sunday to clear the steep slope that led to the lake, build a thirty-foot brick stairway to the water, and plant daffodils. The chance to sit at the table of the most famous woman doctor in the world was an opportunity not to be passed up. Their professional ties, friendship, and Helen's mentorship of Charlotte date to this time. Even at home, her fellows discovered, Helen had her methods, from the way she poured sherry to the way she rang a tiny silver bell at her fingertips during dinner, gently signaling Amy when it was time for the next course.

Informal cocktail hours or suppers were how professors came to know and evaluate their staff. In regularly inviting students to their home, the Parks became like parents to Helen. Sabin, the first woman professor at Hopkins, had a similar relationship with her mentor, the anatomist Mall,

Helen clearing weeds from her new property in Baltimore County with her dog, Spot, ca. 1950. Photo by Charlotte Ferencz.

and his wife. Park's pediatric residents cut his grass on the many Sundays he opened his Baltimore house to them and on a visit to his Cape Breton cottage installed a shower. Sometimes it was the boss who labored to make a subordinate feel welcome; the head of pediatric endocrinology once sodded the lawn of his new researcher's home.

Blalock on weekends socialized with society figures listed in the "Blue Book," people with second homes like his on the private Gibson Island. Occasionally he invited surgical residents. Cooley, who would become best known for implanting the artificial heart, spent a weekend pouring Blalock bourbon and playing pool, earning Blalock's approval early in his residency. William P. Longmire Jr., who would write Blalock's biography, took his family on vacations with the Blalocks. Refusing an invitation was difficult. If he gave a favor, Blalock expected one in return. At Blalock's request, one top surgeon wrote Blalock's wife a lengthy letter describing a trip that she missed. Blalock himself phoned the wives and mothers of his surgeons to boast about them. There was no other group in the medical school as large or with more camaraderie than Blalock's surgeons.

Helen's invitation to perform yardwork implied a quid pro quo that made some fellows uncomfortable. Regulars debated how to take a Sunday off and whether on some Sundays they could show up dressed for dinner and skip the work without offending Helen (yes). Second-year fellows warned newcomers they might be asked for an hour of two of yardwork with dinner invitations. But those who attended developed a special bond. Helen's invitation, unseemly as it was for a boss to ask subordinates to work on her personal needs, was offered with the genuine hope of sharing her love of gardening and to give doctors stuck inside the hospital all week the chance to enjoy the outdoors, get some exercise, and build esprit des corps. It also reflected her simple style; digging in the dirt made everybody equal. She wanted to get to know her fellows, to improve her relationships with them, and if they got some fresh air while helping her execute her garden schemes, so much the better. She always looked for shortcuts.

Perlman, the South African, and other doctors far from home were happy for the chance to socialize. It was a welcome alternative to dining at the hospital on Sundays when spouses of poorly paid interns ate for free. The crowd was international and fun. They took fishing poles down to the lake, paddled about in Helen's canoe, and ice-skated in winter. Charlotte attended almost every Sunday. She had been finishing a fellowship in cardiology at Children's Hospital in Montreal when her chairman returned from Helen's clinic and said she too needed to learn the latest in cardiology in Baltimore.

Charlotte had hardly expected to live through a scientific revolution when she graduated medical school, and now she became an eyewitness. She was also witness to the rise of women in medicine and the links between them. By the time Charlotte arrived in Baltimore in July 1949, Helen and Blalock had already won the most prestigious award in medicine, the Lasker Award, and been nominated for a Nobel Prize. Helen had also captured the public imagination. As the Associated Press's Woman of the Year in science, her photograph was published next to those of Eleanor Roosevelt; the newly elected first woman US senator, Margaret Chase Smith; movie star Loretta Young; and Betty Smith, best-selling author of *A Tree Grows in Brooklyn*.

When Helen returned in September, an intimidated Charlotte stayed quiet in her presence. Helen noticed. Only a month later, Helen impressed Charlotte with her kindness by staging a surprise birthday party for her. By then, Charlotte had noticed something about Helen: the woman who was famous around the world for her contribution to children's hearts was the subject of resentment at home. Outwardly Helen seemed calm; only later would she admit she was very unhappy.

By the late 1950s Helen's working dinners had turned into informal wine-and-cheese drop-ins on Sunday afternoons and, in the evening, buffet dinners of crab cakes and sherry, for alternating groups of fellows. She greeted them outside with a tour of blooming plants, in spring a lawn dotted with forty varieties of daffodils and pink flowering crabapple trees that smelled of roses and, in fall, a trifoliate orange tree with citrus-scented fruits. Then it was on to the greenhouse behind her cottage. There, Helen learned to grow and share plants that brought her much happiness.

*

Only the outlines of this lush landscape—the circular drive with its weeping pink cherry at the entrance, the walking path, and the steps to the lake—were in place on Saturday, April 28, 1951, a clear spring day, when a dozen or so cars began parking in rows on Helen's lawn. Women in dresses and hats and men in spring suits emerged and helped themselves to glasses of wine and beer on their way to gaze at the lake. From garden paths, they congregated around the large brick oven on Helen's back lawn to watch Amy Clark's brother, James Brown, toast crab cakes. Helen had invited current and former fellows over for a social gathering at which she asked them to share their latest research and insights on patients.

Charlotte had flown in for the weekend from Montreal Children's Hospital. In the years ahead, she would take charge of Helen's rheumatic fever patients, publish important papers on the lungs, and conduct the first study to explore the causes of children's heart defects. Other attendees that first year included Whittemore, now at Yale, where she would conduct one of the earliest studies on descendants of children born with defective hearts, finding only 10 percent had heart problems of their own. Mary Allen Engle, who graduated first in her medical school class and had witnessed the historic blue baby operation while on a surgery rotation, arrived from New York City, where she would set up pediatric cardiology at Cornell University. Already, she had a discovery to share: large numbers of babies born with gaping holes in their heart walls were dying of heart failure.[10] Until now, Helen had known only of small, insignificant holes.

The party might have included newcomer Catherine A. Neill, the demure, nurturing daughter of an upper-class British family, who found her way to Helen's clinic in 1951 after earning a medical degree in London and taking a fellowship in Toronto. Tall like Helen, slight of build, and careful in manner, she would return to England after her fellowship only to be recruited back permanently a few years later. As Helen's hearing deteriorated and her role in pediatric cardiology expanded to the international stage, Neill would become the cardiac clinic's mainstay and Helen's right hand. With humor and clarity, she would outshine Helen as a teacher to the next generation of pediatric cardiologists, and author a textbook and history of children's heart medicine. Neill would become Helen's closest female colleague and loyal confidante.

After this gathering, Helen would stage a reunion and an "outdoor barbecue supper" for former fellows every two years for more than a quarter-century. As the number of fellows grew, a discussion of the latest research, treatment, and surgery on children's hearts would evolve into a three-day formal event at Hopkins, kicked off by an afternoon supper at Helen's. She chose the date to coincide with the blooms in her garden. Along with pediatricians specializing in cardiology, prominent doctors in adult cardiology also attended; some years upwards of 150 people poked around her lawn with a view of the lake, drink in hand. Over the years, hundreds of people from around the globe were introduced to Maryland crab cakes. This was how Helen kept her band of children's heart doctors together and formed a new academic field. This was the start of pediatric cardiology as its own new discipline.

*

Professional women by virtue of their small numbers suffered isolation. Many lived together to assuage loneliness, some of them in lesbian relationships. Helen had an occasional dinner with her friend Harriet Guild, who was also single and headed the children's kidney clinic, and a cordial work relationship with the married Caroline Bedell Thomas, tall and aristocratic, who shared Helen's interest in physiology and heart disease. Bedell began one of the first longitudinal studies to examine traits that could predict heart disease. But Helen seemed to have had relatively little social interaction with women doctors whose careers were unfolding at the same time as hers.

Ella Oppenheimer, the married doctor who took over the pathology lab during World War II and became a legendary teacher, regularly crossed paths with Helen at the Parks' home. But Oppenheimer, whose family had deep ties to the Parks, would leave the room when Helen arrived.[11] She had a life outside work and found Helen too serious, uninteresting. Helen may have viewed her as not serious enough, a dilettante. Helen did not help herself when she sent Charlotte sneaking into the pathology lab to collect heart specimens of deceased patients before Oppenheimer could label them. (Helen used them as teaching tools, challenging her fellows to match unmarked specimens to children they had treated.)

Instead, Helen purposively filled her life and her home with patients and their families, visiting scholars, and her fellows and their families. Helen's deepest relationships centered on building her discipline and mentoring the many doctors and students who came through her clinic and built consequential careers. She helped them find jobs, promoted research opportunities, and invited them to sit next to her at professional conferences. She opened her home to them on Saturdays to discuss cases. Marie Brown in the mid-1950s arrived at Helen's door to review the case of a four-year-old patient with tetralogy of Fallot. The surgeon was "hot" to operate using a new method, but Brown, who would follow Whittemore to Yale, wanted the physician and the procedure to mature.[12]

Helen built more personal connections with her fellows too, dining and attending art or music events—in later years, she shared a symphony subscription with Charlotte—and including spouses and children of married fellows at Baltimore barbecues and her weekend guest rotations in Cotuit. She enjoyed watching children run around her bluff, hide in her flower beds, or dig clams on her beach and even invited families to use her cottage when she was not there. (Her fellow Brown stayed months in Cotuit when she was recovering from surgery.) In chatty letters to fellows, Helen asked after relatives and reported on topics from the glow of real candles on her sister's Christmas tree to the Unitarian creed of truth and service she followed. She wrote puns after the poet John Keats (things unheard being sweeter than those heard). On long drives she entertained them with puzzles. In turn, fellows brought Helen a succession of dachshunds, floral silk fabrics for her dresses, and seeds for her garden and kept her informed of the latest medical advances. They styled themselves the Knights of Taussig. The reference to the legendary Knights of the Round Table was fitting, given Helen's view of herself as a leader among equals. If the table was long, she always made sure to sit in the middle because of her poor hearing.

Helen's training program differed from Blalock's in another way. Many of her fellows were women from other countries, and not all were white. Effie Ellis, a highly recommended biologist who specialized in Black premature births and infant care, had to apply twice while the new chief of pediatrics, Schwentker, although supportive, sought approval from the hospital board to hire a Black doctor. (Hopkins had only ever employed one part-time Black physician, and adult patients remained segregated by race until the late 1950s.)[13] When Ellis arrived in 1952 from a pediatric residency at Massachusetts General Hospital, another doctor confronted Helen, asking where Ellis would eat. "The dining room," Helen replied. The physicians' dining room was small, reflecting their numbers. The doctor said he would stop eating there.[14]

Helen's fellows in the early years included men and women who trained at top medical schools in Great Britain, Ireland, Canada, Europe, and the Middle East. Later they came from Asia and Africa. By the mid-1950s her staff regularly included four Americans and four foreign fellows, often from countries with poor health systems for women and children on training grants from the National Institutes of Health.[15] Some stayed, but most returned home to start or improve children's heart programs and interest surgeons in their patients. They included Kamala I. Vytilingam who started the cardiology program at the Christian Medical College and Hospital in Vellore, India, and the first chair of cardiology in Germany, Alois Beuren at the University of Göttingen. These connections would prove critical as Helen sought to protect mothers and children from new and dangerous drugs.

17 A CHESS GAME

Initially, Helen recommended surgery using a simple formula: if the patient was near death and the operation offered a chance at life, Helen recommended it. But as operations matured and Blalock introduced new ones, Helen found herself assessing whether patients who were not at death's door would live longer if operated on now, later, or not at all. She and Blalock saw quickly that survival rates differed by age. A child older than three, whose heart was not yet enlarged, was the ideal candidate. Teenagers fared poorly. Multiple other factors were in play: the condition of the patient, the degree of deformity, how long Helen estimated patients could live compared with the newness of an operation, and how individual conditions could change over time. It was like a chess game with a bewildering choice of strategies.

Finally, Helen's assessment of risk was affected by her view of the dangers of anesthesia, the drugs used to numb and slow down body processes during surgery. Her patients already suffered from dangerously low oxygen levels, and slowing their blood flow further put them at extreme risk. Deaths related to anesthesia occurred at every hospital, and without data, it is impossible to compare rates in these early years. But in Blalock's operating room, the delivery of this crucial drug did not improve as quickly as it did at other first-tier hospitals. For fifteen years, Blalock insisted that his hospital-trained nurses deliver anesthesia, though other doctors believed it should be delivered by research-oriented doctors with special medical training. One after another, these doctors refused to work with him. Austin Lamont, the anesthesiologist who had declined to provide an antibiotic for the first blue baby on grounds it might kill her, quit when Blalock refused to let him develop a separate department and introduce residential internships like those in surgery and pediatric cardiology. (The exam for this specialty began in 1939.) Lamont's mentee, Merel Harmel, quit a year later, in 1947, after Blalock blamed him for the death of a baby anesthetized by Blalock's favorite nurse anesthesiologist, Olive Berger, with the drug cyclopropane. The nurse had asked Harmel to oversee her. Blalock complained the drug was changed without his knowledge.[1] Cyclopropane sedated a patient quickly, but too much of it, too rapidly, set off cardiac arrest. There were none of the usual early signs—changes in breathing or drooling at the mouth that nurses relied on to detect trouble and change the dose. The lack of clear lines of authority was troubling. Little changed over the next five years except that the equipment to supply patients with drugs and oxygen aged and grew less efficient.[2] The concentration of oxygen in tents was hardly better than in the air outside them.

In this fragile environment, Helen quickly decided that a patient who could wait for surgery should. In her view, a child with an obstructed aorta but a normal size heart and no problem exercising his arms, legs, and voice was far safer appearing at her office once a month so she could monitor the size of his heart than undergoing surgery. This wait-and-watch strategy could last years.[3] Because tools and techniques improved so rapidly, she reasoned, even a month's delay could mean the difference between life and death. Helen's conservative approach required systematic review of surgical outcomes and many more patient exams and blood tests. She hoped the huge amount of data she collected would tell her how long patients could wait, whether and why and when surgery worked, and when it might prove fatal. Instead of worrying about fitting patients into the operating room schedule, Helen now worried about how much blood the lungs of a patient could handle, how much oxygen they could make available at what stage of their lives, and why some vessels carried blood to the body more efficiently than others. She was looking for patterns.

Her decisions were so complex that Helen was forced to consider yet another risk factor, her own state of mind. One day, faced with wrenching decisions on multiple patients, Helen suddenly got up from her desk, put on her hat and coat, and announced to her staff she was going home. Some families in the waiting room outside had already checked out of their hotels and planned to leave later that day. One father had plane tickets in his hand. Helen walked out of her office, apologized profusely for her early departure, and asked the families to return the next day, when her mind would be fresh. She feared her exhaustion would affect her judgment. "I want to make a good decision for you," she told them. "I want to give you my best thoughts." Helen's fellows braced themselves; they expected parents to be upset. Instead, the parents seemed glad. One after the next rose to thank her. It was as if Helen's exhaustion had revealed to them how fully she burdened herself with decisions about their children. Incidents like this explain why they returned to Helen for advice again and again.

*

At the very time Helen was developing her conservative formula for recommending surgery, the number of blue babies operated on at Hopkins declined dramatically. In two years, from the peak year of 1947 when two hundred operations were performed and the end of 1949, the number of blue baby operations dropped by half. Most of the drop was probably due to competition from medical centers in major US cities—at least seventeen hospitals had opened cardiac surgery programs—and around the world. The flood of patients from France slowed to a trickle once Paris hospitals began the operations, and in London, doctors had operated almost immediately after Blalock's breakthrough. Some of the decline in volume may have been due to Helen's increased caution, though we don't know whether the rate at which she recommended patients changed. By 1949 she had reason to be alarmed: for the first time since blue baby operations began, the mortality rate rose, to 8.9 percent.[4] Two simpler operations performed with regularity, clearing a clogged aorta (coarctation) and closing an open fetal duct, could not make up for the decline in blue baby cases, even if Helen had recommended them at the usual rate. But young surgeons attracted to Blalock's training program needed patients to develop their skills and advance this new medical frontier. While other types of surgeries continued, the shortage of candidates for heart surgery exposed a brewing conflict.

Blalock often deferred to Helen and shared her view of when or whether to operate and her idea that patients should be monitored closely for changes such as an increase in heart size that would shift the balance in favor of an operation. But from at least 1948, Blalock's surgeons found themselves at odds with Helen's recommendations. Blalock and his team, too, reviewed outcomes and worried about long-term success. But their risk

assessment frequently differed from Helen's. They resented being dependent on her recommendations for what many now considered routine
operations. "Who is *she*?" some surgeons wondered. Why was Helen or her
fellow always around the operating room and recovery suites? And why
was Blalock putting up with her? In the early days of blue baby operations,
Blalock would tell interns rotating through surgery there was no one else
in the world who could diagnose the condition they were seeing.[5] He also
acknowledged that Helen was the one who suggested the concept of the
blue baby operation.[6] Blalock respected her intellect as well as her ingenuity
and resourcefulness, even if her methods annoyed him. "He trusted her,"
Richard S. Ross, a chief medical resident in Helen's era and later the Hopkins
medical school dean, recalled. Helen "would be in with a patient explaining
a case and Blalock would come in and before she finished, he would say 'ok
put him on the schedule.' She felt a little outgunned," he said.

But in 1949 surgeons working with Blalock had little inkling of Helen's
studies of defective hearts, the samples she had examined at autopsies,
or that she had lobbied Blalock and Gross before him to create a vascular
shunt for her patients. It was all the more mysterious since Blalock had a
lock on every other aspect of his department. He deputized chief residents
and rewarded loyalty with jobs or recommendations. He beat back efforts
to develop specialties outside his control.

Unlike others Blalock tried to overpower, Helen did not back down or
go away. The strategies he had tried, writing papers without her, inserting
deputies as go-betweens, had not deterred Helen. Helen had gone on to
write her own papers. She was directing her fellows to develop papers on
aspects of surgery, too, and interesting some of his surgeons in her clinical findings and solutions. Helen controlled the critical flow of patients;
her patients clung to her and she to them. They included chronically ill
rheumatic fever patients who had grown into adults with failing hearts
and new adult patients attracted by her reputation. They came to Hopkins
because of her. The hopeless ones, most of all, were attracted to Helen. And
these were the ones she would be most likely to recommend for Blalock's
experimental surgery.

To prod her into sending him more patients, Blalock told Helen about
Christmas cards he received from children with reversed blood flow in the
major arteries (transposed major arteries) in whom he had created openings in the wall between the upper heart chambers to improve circulation

of oxygenated blood. "Since we have not had candidates in recent months for this procedure," he wrote Helen in early 1950, "I simply wish to remind you of these moderately encouraging results."[7]

*

Patience was not a trait shared by sharp-edged surgeons leading a medical revolution. By late 1949, the year of the steep drop in blue baby operations, Blalock was "having a great deal of difficulty" with Helen because she refused to recommend operations in cases where "to most of us the patient seemed to merit an operative procedure,"[8] according to William Muller, then the chief surgical resident. During this "rather slack period" for surgery, Muller decided to confront Helen. An opportunity arose in her clinic one day when he observed her examine a patient with a narrowing of the aorta. Helen told the patient he could wait and recommended against surgery. To Muller, "there was no obvious reason for this patient to wait because coarctations had been operated upon very successfully for several years."[9] Muller openly questioned Helen: Why had she refused to recommend an operation? That Helen knew he was being impertinent can be inferred by what happened next. Although she reveled in giving explanations, she offered none. Instead, she told Muller that if he did not like her recommendation, he should leave her office. Out he went, to find his boss.

Blalock dissolved in fury as he listened to Muller describe his abrupt dismissal. The chief surgical resident was a stand-in for him, and he commanded the most important department at Hopkins; Helen, whatever her expertise, was subordinate to both. Twice before Blalock had threatened to stop working with Helen, but now her tortuous decision-making affected his operating room schedule and surgical training program. Blalock vowed to open a cardiology service under his own command. With Muller at his side, he marched downstairs to seek approval from the director of the hospital, Edwin L. Crosby, to open a division of cardiology services, presumably for adults, within the department of surgery. This new division would be under Blalock's control.

*

In the months leading up to this confrontation, Helen and Blalock had had an unpleasant exchange, in which Helen had dared to bring up her old complaint that she had not received fair billing on their original blue baby paper. She had not mentioned it for four years. But in the summer of 1949, irked by what she saw as another attempt to minimize her contributions, she could not resist renewing her demand for fairness.

The trigger was Blalock's draft paper announcing pioneering surgery to help children with transposed major arteries. It followed a year of experiments on dogs and at least a year of operations on Helen's patients. Helen had been the one to identify and separate children with transposed arteries from typical blue babies. Besides a breakthrough 1938 paper on diagnosing the condition in the living, she had written a definitive fifty-page description of nine variations in her book and explained how additional defects enabled these children to live. Beginning in 1948 Helen had supplied Blalock with patients for the operation, many of them very small and very sick, and Blalock lost twenty-two out of thirty-three patients trying three different methods to help them.

In directing Vivien Thomas in the laboratory, Blalock had asked the same question about patients with transposed vessels that Helen asked about tetralogy of Fallot blue babies: Why did some children live longer than others, how were they different, and what could be done surgically to replicate whatever kept the older children alive? Then he looked for clues. From autopsies, Blalock and his resident, C. Rollins Hanlon, realized children who lived longer had one or sometimes two holes in the heart wall, so that a small amount of oxygenated blood from the lungs could spill over and mix with blood going to the body. Blalock reasoned that some children would live longer if he could create a second hole or additional passageway, so more oxygenated blood could get to the body. Vivien Thomas's solution, a new hole between the pulmonary vein and the upper right heart chamber, the atria, captured oxygenated blood as it looped again and again through the lungs. This higher oxygen blood then moved from right to left heart and into the body. Blalock was stunned when he saw it, telling Vivien Thomas "this looks like something the Lord made."[10] Hanlon refined the surgery, ensuring the hole remained open. This was the first operation to extend the lives of children with transposed aorta and pulmonary arteries.

The draft Blalock sent Helen for review cited her book and credited her for the diagnoses. His discussion of this important surgical procedure was "of course, strong!" she told him. But his description of the problem was weak. She proposed to add an introduction and discussion by her to strengthen it. It was the same organizing structure he rejected for their blue baby paper when he insisted on his name first.

A surgery-only paper was Blalock's prerogative—Helen acknowledged that. But this one involved an important operation for a group of babies on which Helen was the expert, and she had learned even more about their condition since her book was published. Her tone was professional, but her critique pointed. Bing's theory about blood flow, which Blalock cited, that one side of the circulation would be depleted of blood without a balance between two sides? It never happened in practice. The photo he planned to use? The defect was in the wrong place. A sketch would be easier on the reader, and she drew it on the back of his paper. Helen next turned to Blalock's surgical technique. Could he possibly try grafting the two vessels at a different point? She provided details. Would he consider returning to the lab for additional experiments?

This would have stung, but not as much as Helen's decision to complain once again about attribution in their original 1945 paper. "Although we have both receive great credit from that paper, I believe that it would have been fairer if that had been by you with an introduction by me instead of my name second," she wrote Blalock.[11] In a terse, one-paragraph response a week later, Blalock said, no, he was not in favor of an introduction by her. She returned his draft, marked up "unmercifully" in her own words. Fine, she told him, there was no need to have her name on a mostly surgery paper. (Helen was not the only one who felt slighted in this important paper; the architect, Vivien Thomas, would not be recognized until he wrote his account of this surgery.)[12] This was the low point in their two-decades-long relationship. Within months of this exchange, Blalock moved to set up a clinical cardiology service of his own.

*

By now, around the time of the one-thousandth blue baby operation in 1950, Blalock was revered not only for his path-breaking surgery but also for restoring Hopkins's reputation as the leading US hospital for surgery and the model eight-year training program for surgeons, a reputation established at the turn of the twentieth century by the European-trained Sir William Halsted. Surgery was the blockbuster solution of the day for medical ailments. Happy patients were making significant donations. Blalock was intricately involved in planning a new surgery center and laboratory. His department funded other programs whose work complemented his own.

The residents Blalock attracted were looking at illustrious careers. One had already named a son after him, and he was a godfather to others. They recognized that his one-thousandth operation on a blue baby was a champagne moment, something to be memorialized.

To honor the doctor in whose shadow they worked, Blalock's residents hired a noted photographer, Yousef Karsh, to make a memorable image of "The Professor" as they sometimes called him. Karsh was a star of the postwar art world, one of the 1950s artists experimenting with realism. The black-and-white photo images he made of Blalock were a break from the formal oil paintings that adorned the walls of academic institutions. They telegraphed smart, powerful, and confident, as if his achievements had been easy. Blalock's residents called an impromptu gathering in his living room to select the ones they liked and toast their boss.

If Helen had been invited, she would have admired the images. She had little use for formal portraits. The Karsh photo was authentic and would become iconic. Did Helen feel a tinge of longing for a photo of equal import? It was a lot to expect at this stage. Beginning with only a few doctors training in her clinic, serving two-year fellowships compared to six or eight years for a surgical resident, Helen did not yet have a large, cohesive group of followers to underwrite a portrait. Delayed by her book and her patients, she had only begun to build a community. Her effort to get to know her fellows had been complicated in the early years of the blue baby operation as she concentrated on before- and after-care, collected data, and searched for money for staff and patients. And there was the work: for Blalock to operate on one thousand children with tetralogy of Fallot, Helen in the early years would have had to examine three thousand candidates.

Curiously, there is a photograph of Helen taken around this time, in 1949, and it is the most stunning portrait of her found in her personal papers. Little is known about who arranged for it or the New York photographer, who signed his name Jean Duval, or whether it was meant to hang on a wall. It is classical but gently so. Her head slightly tilted, her pose noble, it reveals a wise woman adorned in pearls.

*

When Helen needed to calm her fury, she would take off her glasses and walk up and down in her office, thinking, often with her lips pursed. She walked constantly in 1949, a year more difficult than the first year of blue baby operations and without the happy outcomes that made it bearable.

She had reignited her fight over due credit with Blalock. She avoided Bing, who continued to try to lure her into a battle of semantics, insisting that her original idea to help blue babies by increasing blood to their lungs was incorrect. Her new boss, Schwentker, was trying to outflank her with rheumatic fever research outside her control. And now Blalock was proposing his own cardiology division. These battles coincided with an unexpected spike in blue baby mortality rates and less-than-satisfying surgical solutions for babies with transposed great arteries.

There was more: animal rights protesters had snuck into Blalock's research lab and taken photos of dogs in cages. Now they demanded that Baltimore City stop depositing abandoned dogs at laboratories at Hopkins and the University of Maryland. The cutoff threatened animal experiments that could produce more lifesaving procedures. In the spring, Helen had begun taking some of her patients to meet Anna, a retired, tan-haired mutt who had survived an experimental blue baby operation and was now a pet in Blalock's lab, and staging newspaper photo opportunities.[13] By fall, she found herself calling former patients and organizing prominent doctors to testify against a proposed ban at Baltimore City Hall. Voters would defeat a referendum to protect dogs.

Of all these crises, Blalock's attempt to set up his own cardiology division would be the most consequential. Whether or under what circumstances Helen spoke with hospital administrators is unknown, but she had allies who would have swiftly countered Blalock's attempt to wrest control of patients. Blalock's proposed division of cardiology further concentrated his power, to the consternation of some at Hopkins who felt the balance was skewed.

The impact of Blalock's proposal on patients rather than concerns over his growing power may have doomed it. A cardiology division under the control of the surgery department raised the appearance of a conflict of interest: if the patient had an operable condition, surgeons might be more inclined to operate. As important, A. McGehee Harvey, head of the department of medicine, the other major power center at the hospital, opposed specialization of doctors in cardiology. A master diagnostician like Helen, Harvey oversaw physicians who treated adults for all conditions, including diseased or aging hearts. After the first blue baby operations, these "internal medicine" doctors spent months in Helen's clinic to learn about heart defects and surgical solutions. Harvey was unlikely to accept Blalock's

command over doctors on his staff or allow surgeons to overrule his doctors' recommendations.

The operation that sparked the confrontation had indeed been successful. Blalock or his surgeons had performed at least thirty coarctation operations, with four deaths. What part of this success could be due to Helen's
careful selection of patients? Had she saved lives by delaying surgery? The
answers were unknowable; there was little information but for what was in
Helen's mind.

There were data on declining survival rates for tetralogy of Fallot patients.
Could that have influenced discussions at the top echelon of the hospital
over Blalock's demand for his own cardiology service? What about success
rates for other so-called routine surgeries like the blocked aorta? One death
in particular underscored the risk of routine surgery and raised questions
about the delivery of anesthesia, which in this period was run by nurse
Berger and supervised by Blalock himself. It was a child who received a dose
of cyclopropane as preparation for surgery to close her open fetal duct. Her
heart began to beat rapidly and develop spasmodic contractions, which
led to her death.[14] The surgeon, Cooley, only later learned that cyclopropane could cause irregular heartbeats. Cooley was upset enough by the little
girl's death that in the following months, he built a makeshift version of a
gadget he had heard was under experiment by others on campus to revive
patients in the operating room. It sent an electrical shock to the heart to
reset its normal rhythms and would be used in the operating room for the
next decade.

According to Cooley, Blalock became interested in his project only after
reading Cooley's paper about it in 1950.[15] Afterward, Blalock welcomed into
his own lab a Hopkins engineer working on a project to reset heart rhythms.
For over a dozen years beginning in the late 1930s, Hopkins engineering
dean William Kouwenhoven had been experimenting in the laboratory
with ways to revive the hearts of electric power company workers struck by
lightning. Kouwenhoven, with suggestions from Blalock and others, would
develop a history-making machine to reset and even restart the heart when
it went askew, making surgery safer. It was the defibrillator.[16]

Blalock next shored up the delivery of anesthesia. He still insisted on
hospital-trained nurses, but in 1951, he hired an otolaryngologist, an
expert in blockages in the ear, nose, and throat. Donald F. Proctor, after a
year's training in anesthesiology at the University of Pennsylvania, made

it his business to hire more doctors and get them certified in anesthesiology. Beginning in fall 1952, he modernized equipment and oxygen therapy and improved nurse-anesthetist training. Proctor also systematically
tracked drugs administered before anesthesia and during and after surgery.
It was apparent to him that a major cause of mortality was flawed choice
of preanesthetics, anesthetic drugs themselves, and procedures. Among
twenty-five deaths attributed in part to anesthesia, fifteen were related to
an "improper choice" of anesthetic or method of delivery. Proctor insti-
tuted new protocols and told his staff they were duty-bound to challenge
surgeons who made faulty choices.[17]

An adult heart specialty department with its own chief was inevitable
given the rapid growth of research and the implications of Helen's work
for aging adults who developed obstructed arteries, failed valves, and other
conditions like the defects she diagnosed in newborns. This would be a big
business, but Blalock would not oversee it.

*

In response to what she observed, Helen in the 1950s wedged herself ever
more tightly between patients and those who would operate. The problem was success itself. Surgeons were eager to try out their new techniques.
Patients who once feared the surgeon's scalpel read about life-changing
outcomes and begged for operations. Deaths, still routine, rarely made the
newspapers, but Helen saw them regularly.

In 1952, with blue baby surgery well established and surgeons tinkering
with other new operations, Helen laid out the changing landscape and her
principles for recommending patients for surgery in an article published
in the *Journal of Pediatrics*. It is remarkable in that its goal was to influence practitioners' attitudes and behavior based on her clinical observations, before definitive data on surgery outcomes could be analyzed and
published. Its tone reflected Helen's extreme worry that children would die
from brash decision-making. She set out what amounted to an ethics code
for pediatricians who confronted decisions for the first time about risky
cardiac surgery. She urged them to equip themselves to stand firm against
surgeons and patients.

Referring doctors—pediatricians newly specializing in children's hearts
and those learning on the job at the dawn of this era—had new responsibility not only to get the diagnosis right (they could read her book, for starters),

she wrote, but also "in the selection of patients for operation and *in the prevention of needless operations* [emphasis added]."[18] Saying patients could live years with their defects, depending on the type and severity, Helen warned against liberal referrals or false advice to patients that they would die without operations. Doctors should recommend surgery for patients who were slightly incapacitated when there is "little risk and great benefit," Helen said. Operations on infants were justified only if the infant was about to die, since those with troubled breathing usually survived for some time. On the other hand, a doctor should recommend a risky operation for a patient near death even if the benefit is marginal. "When hope is near," she wrote, "await its safe development." When hope is distant, doctors could help mothers realize that despite a physical handicap, many individuals "can lead happy significant lives and take their places as respected, useful citizens of the world."

In her article, Helen introduced an idea that over the decades would become her mantra: the brilliant light shone on successful surgery hid the "dark sorrow" of its failures. This was personal. She sent many, many babies to Blalock in these first years. Some died because they were too sick. Some died because they were misdiagnosed. Her own mistakes imprinted themselves on her mind, she liked to say, and she did not want others to repeat them.

*

As she developed principles for recommending surgery, Helen applied them to herself. At stake was whether she could continue to provide the best care for her patients. The work that made her famous was accomplished while she coped with what she described as poor hearing. But in the six years since, her hearing had worsened substantially. Records kept by her lipreading tutor reveal she suffered from otosclerosis, in which one of the bones of the middle ear develops abnormally and cannot amplify sound. This is often an inherited condition; both Helen and her father were bothered by extraneous noise that interfered with concentration. It can also occur from bacterial infection, such as the whooping cough Helen contracted in 1929.

By 1951 when Helen approached an expert about her problem, surgeons had a decade's experience carving a tiny path around the bone to let in sound, a procedure called fenestration. It did not always work. Sometimes the bone grew back, closing the hole. After consultation with a hearing specialist, associate professor William G. Hardy, Helen rejected surgery, but she did so only after assuring herself that her poor hearing would not interfere with her ability to be a good doctor. Clearly it did not affect her diagnoses. By the 1950s, she could diagnose a heart defect by the way a child walked or was wheeled into her office, from his build, his age, his color, his energy.[19] Her hands and other tools merely confirmed it. But to understand and comfort patients—especially dying patients whose lip movement slowed—and communicate with colleagues as her hearing declined, Helen developed elaborate strategies. She paid more attention to faces instead of lips, looking for context. She orchestrated one-on-one conversations in her home and the doctors' dining room; it was hard to follow Blalock's lips in his office, where he smoked. By 1951 she had installed amplifiers on phones in her office and in Cotuit. Her hearing aids fit into her ear instead of hanging from a string around her neck. (Her ears were back up where they belonged, she told people.) At home, to alert her to noises of people or cars, she relied on Spot and his successors. Helen's dogs sat on her lap when she wrote at her desk, the excuse she used for sloppy penmanship in letters to friends. Because of her hearing and to remain focused on patients, Helen delegated secondary speaking roles. Although she accompanied students and residents on rounds, Neill ran the clinic from the mid-1950s onward, taught students to use the fluoroscope, presided over patient conferences where fellows analyzed their findings, and summed up their cases for Helen.

Her voice sounded tinny and uneven; she could not hear herself to adjust the volume, and her voice rose and fell without notice. She often skipped or elided words. Remarkably, descriptions of Helen by patients, families, and news reporters mention her height, the color of her hair, the curiosity in her eyes, the style of her suit, her "large, capable hands," as one reporter wrote, but not her deficient hearing.

"People were very aware of it, but also very kind to her," Patsy Perlman recalled. "You had to pay close attention, and she demanded that you pay close attention." By the end of the decade, Helen's office assistants had to repeat themselves, but she communicated unimpeded with patients. The work that ensured her legacy, when she would develop a medical specialty and emerge as an international advocate for patients, was undertaken in years when she had very poor hearing.[20]

18 THE END OF RHEUMATIC FEVER

At the height of her fame, when blue babies rushed her clinic and she arrived home each night weary from their care, Helen never lost sight of her rheumatic fever patients. "The 'blue baby' operative technique is dramatic," she told a *Sun* reporter in 1946, "but the need for that technique is small as compared with the treatment needed for thousands of children and young adults who are victims of rheumatic fever, which often carries with it a cardiac weakness unless treatment is carefully followed." She pleaded publicly for more convalescent beds, attended luncheons to woo potential donors, and continued to visit her rheumatic patients weekly to be sure they followed her orders for bed rest.[1] When she detected the sound of a leaky valve, Helen remained adamant that children just past the acute stage of the disease continue to lie still, sitting up only to eat from a little table next to the bed; she wanted their hearts to be able to repair themselves.

But rheumatic fever was less virulent in the 1950s for reasons no one understood—even when Helen visited, the nuns at St. Gabriel's could not keep children in bed—and the number of rheumatic fever patients was waning. How the disease worked was still unclear, although the foundational laboratory work on the bacteria behind rheumatic fever was already done. Rebecca Lancefield, a microbiologist at the Rockefeller Institute, had subdivided and labeled strains of strep bacteria into groups, putting strains from human diseases in Group A in the 1930s, and now was working to further understand the composition of Group A streptococcal bacteria she identified as responsible for diseases like rheumatic fever.[2] Helen's friend from medical school, Kuttner, head researcher at a large convalescent center in

suburban New York, Irvington House, had reported in 1946 that despite
a massive outbreak of strep in her facility, no patient was reinfected with
rheumatic fever, meaning that not all strains of Group A streptococcal bacteria caused strep throat. Doctors were working to identify the precise strain
that caused the disease.

The next question would be whether and how much of the new antibiotic, penicillin, could stop it. An expert in immunology with experience in
rheumatic and scarlet fevers, the chief of pediatrics, Schwentker, had been
a virologist at the Rockefeller Institute after graduating from Hopkins and
working under Park. Before he returned in 1946, he announced to New
York newspapers that he would conduct a new long-term study of rheumatic fever to test penicillin. Although his academic interests paralleled
Helen's, Schwentker did not like what he saw happening in her clinic and
wanted one of his own.[3] The reason is unclear, though we know he was
interested in laboratory research while Helen was pleading for beds to still
the inflamed heart. She had begun to use penicillin but not to test it. It was
just becoming available.

"Absolutely no, absolutely no," Helen responded, when he sought his
own clinic.[4] In addition to her experience with this disease, her clinic was
underwriting some of the pediatrics department. Helen's fame meant she
attracted patients, a situation any boss would be careful not to upset. Her
grants to study the treatment of rheumatic fever patients, which helped
fund fellows' salaries, would dry up if she ceded patients. Instead, she
invited his pediatric fellow, Milton Markowitz, who had worked under
Kuttner at Irvington House, to treat and study patients in her clinic.

Schwentker found ways around her. He assigned Markowitz to Helen's
clinic two days a week, but also had him volunteer to coordinate care at
Happy Hills, a thirty-bed Baltimore home for rheumatic fever patients established in the 1920s, and eventually, after Markowitz left for private practice,
helped him get his own clinic. Schwentker achieved this with funds from
a new rheumatic fever association in Maryland that he ran for four years.
This association, later the Maryland Heart Association, also partly funded
new independent rheumatic fever clinics in the city outside Helen's control. Schwentker got Markowitz appointed to the association's board.

Loud and unpredictable, Schwentker was a stark contrast to Park. Helen
managed because she enjoyed her day-to-day work and because building
a house of her own had given her a "certain self-confidence" she had not

expected. "I'm happy now & I can meet Francis halfway with a smile and everything is easier," she wrote to Charlotte.[5] She won Schwentker's support in battles with Blalock and to open a heart clinic in Martinsburg, West Virginia, so poor patients would not have to travel to Baltimore.

*

While Helen was busy advocating for one hundred more convalescent beds to prevent a recurrence of rheumatic fever, Markowitz devised a more practical way to prevent patients from relapsing. More important, he devised a way to prevent them from developing the disease in the first place. Shortly after his pediatric residency in 1949, he joined the city's leading pediatrician, Alexander J. Schaffer, to form the first group pediatric practice in Baltimore and one of the earliest nationwide. This alliance was important because of the volume of patients they treated and because these doctors were academics at heart. They had gone into private practice to pay their bills, but they observed, reported, and taught what they learned from their large patient base. Markowitz alone saw a thousand patients a year in their homes and at his central city clinic. He continued his twice-weekly work in Helen's hospital clinic and oversaw care at Happy Hills, the private convalescent home. He was positioned to act when he learned from a preliminary report in *JAMA* in 1950 that Lewis W. Wannamaker and associates at a Wyoming Air Force base had prevented rheumatic fever in children by identifying patients with sore throats caused by the streptococcal pharyngitis (Group A) bacteria and immediately treating them with penicillin.[6] Unlike the sulfa class of antibiotics previously used to treat rheumatic fever, which only reduced inflammation, penicillin killed the strep bacteria.

Markowitz now paid careful attention to patients who came to him with sore throats. He diagnosed streptococcal pharyngitis infections by culturing them in his office rather than send samples to the hospital. This allowed him to treat the child with penicillin within a day or two instead of a week. He talked about his method on rounds with doctors training at Hopkins and published an article about it, including how to diagnose strep throat. City doctors adopted his method, even ordering throat culture kits like the one he had asked a local company to make. Soon these kits were in the hands of doctors throughout the country.

Helen followed his research and adopted his methods to prevent recurring disease and continued to monitor patients with compromised hearts.

Her responsibilities were so large now that she needed a deputy. Charlotte, meanwhile, needed a job; she had lost hope of building her medical career near her family in Montreal. Despite her credentials, the head of cardiology at McGill made her feel unwelcome after her residency. The country's other top cardiologist, John Keith in Toronto, refused to hire a woman. When Helen pushed to hire Charlotte using one of the medical school's coveted slots for visa sponsorship, Schwentker agreed on one condition: that Helen make Charlotte the full-time director of a separate division of rheumatic fever, although the division could remain under Helen's clinic.

Charlotte's return in 1954 brightened Helen's outlook. She knew her rheumatic fever clinic needed a boost, she told Charlotte. She had also hired Catherine Neill to help train fellows, so she had two full-time doctors to manage key aspects of her clinic. (That first summer, she even offered to rent them her house, minus the silver she stored in the bank and the rugs she took up for cleaning.) The rheumatic fever clinic remained separate, and penicillin studies continued, with Charlotte in charge, even after Schwentker's death by suicide within months of her arrival. But with fewer rheumatic patients, Helen invited Charlotte to collaborate with her then-senior fellow, Frank Dammann Jr., to investigate the impact of new heart surgeries on the lungs. Eventually Helen helped Charlotte land a job with a favorite former fellow, Edward C. Lambert, who was conducting advanced research into childhood cardiology problems at the State University of New York Children's Hospital.

By 1958, the year Charlotte left to become an assistant professor in Buffalo, Markowitz had begun steering all first-time rheumatic fever patients to his own clinic, for his studies on preventing recurrences, leaving Helen to treat patients already exhibiting heart damage. Within the year he convinced his mentor, Kuttner, to return to Baltimore to work with him. With grants from the National Institutes of Health and operating out of laboratories at Baltimore's Sinai Hospital, they conducted the single most important study on preventing recurrence of the disease.[7] Their 1965 paper set the international standard for prevention of rheumatic fever and contributed to the worldwide decline in the disease.

Accurate identification and rapid treatment of strep throat by pediatricians in the community was a major reason for the elimination of rheumatic fever. The decline in the ferocity of the disease and less crowded living conditions that checked the spread of bacteria before the penicillin era were also factors. But once doctors began identifying and treating strep throat, the disease all but disappeared. Between 1960 and 1980, the incidence of initial cases of rheumatic fever in Baltimore dropped from about 40 per 100,000 people to 0.05 per 100,000.[8] There were still funerals, but less than fifteen years after Charlotte began accompanying Helen to St. Gabriel's convalescent home, it would be shuttered.

Park had never answered Helen when in 1936, nearing a breakdown, she had demanded to know whether the purpose of her clinic was to treat rheumatic fever patients or study the disease. The answer was now clear. Just as Park and others at the beginning of the twentieth century had hoped, treating children with chronic disease would ease their suffering and lead to discoveries of how it worked and ways to prevent it. Rheumatic fever was now entirely preventable.

Significantly, the collaboration of Schwentker, Markowitz, and colleagues, including clinics they funded throughout the city, was one of the last such arrangements between academic medicine and community-based doctors. Medical schools hired full-time faculty to conduct research into other chronic conditions that would involve multiple academic medical centers across the country. This left doctors in private practice who depended on patient fees to provide people with basic care for preventable conditions.

19 ON HER OWN AGAIN

Heart surgery began on children but quickly expanded to adults, who represented a far bigger customer base for physicians and surgeons than children with rare birth defects.

Adult heart problems stemmed mostly from narrowing of the coronary arteries caused by diet and age. Aging adults developed the same thickened or otherwise faulty valves acquired from rhematic fever disease or birth defects, and internal medicine residents studied in Helen's clinic to learn to diagnose heart defects so they too could recommend patients for surgeries. Before, the only formal training for adult cardiology was to analyze EKGs in the laboratory.

By 1953, the chief medical resident, Ross, convinced the chair of medicine, Harvey, to allow him to specialize in the heart. This decision encouraged other medical residents to specialize, including by studying in Helen's clinic, and helped speed the growth of adult cardiology and nonsurgical services and technology at Hopkins and elsewhere, notably the new National Heart Institute (now the National Heart, Lung, and Blood Institute of the National Institutes of Health). When mitral valve surgery on Helen's adult rheumatic patients became possible in 1949, her friend E. Cowles Andrus, a European-trained expert in cardiac arrhythmia (irregular heartbeats) who oversaw the heart laboratory, helped Blalock develop rules for selecting patients.

But for five years as adult heart treatment options multiplied, Helen refused to give up her adult rheumatic patients, many of whom she had known for decades. If her pediatric fellows found themselves ill equipped to provide nonsurgical treatment for adult patients, she had them consult medical residents for guidance. If surgeons assigned Helen's patients to adult cardiologists for postoperative care, they faced her reprimand. Succeeding Ross as chief medical resident was a doctor with similar patient care instincts, Sherman Mellinkoff, who would become the founding dean of a medical school at the University of California, Los Angeles. As Helen got to know the young doctors treating adult heart patients, she grew increasingly reassured that they could provide better care than her own fellows. In 1954, she was convinced to sign off her adult patients to Ross, becoming his consultant. Once adults were admitted into the hospital, they would be his. A department for adult cardiovascular medicine opened a few years later, in 1957.

Helen still sneaked her adult rheumatic fever patients into her clinic for an exam on occasion and continued to advise adult patients with heart defects on whether to undergo a second surgery. She was nearly eighty when she accompanied one former patient back into the operating room. Her studies on the quality of their lives lasted decades, and she never stopped answering their letters.

*

Meanwhile, new surgeries for the heart proliferated. After the blue baby operation, Blalock and a handful of other surgeons dared to move closer to the pumping action itself, to repair the valves that opened and closed the chambers that rhythmically moved blood to the body and lungs. The earliest experiments focused on the mitral valve, which controls blood flow from the top to the bottom of the left heart and is the first to be attacked by rheumatic disease. To open mitral valves that had grown hard from disease or had not worked from the start, the surgeon's only option was to reach into the beating heart muscle with his finger to feel for the valve and then cut through its thickened wall. The goal was to create a space big enough to restore adequate blood flow without interfering with the heart pump. It was chancy; the surgeon could not see the valve.

Horace Smithy, a gentlemanly South Carolinian, performed the first successful mitral valve operation in January 1948.[1] Five of his seven patients survived, but Smithy was too busy trying a far riskier operation to publish his results.[2] He wanted to repair the aortic valve, which allows the left heart chamber to push blood into the aorta, the blood vessel that distributes blood throughout the body. Fixing aortic valves was riskier than fixing mitral valves. An obstructed or narrowed section of aortic valve (aortic stenosis) was harder to reach and more sensitive than the mitral valve because of its role in moving blood into the body. Unlike Helen's tetralogy of Fallot patients, patients with a damaged aortic valve had no backup paths for blood flow. Smithy had worked on it in his animal laboratory far longer than he had worked on his mitral valve operation, partly out of self-interest: he had a damaged aortic valve himself as the result of childhood rheumatic fever and was experiencing heart failure. He hoped that if he taught his operation to Blalock, Blalock would perform it on him. Blalock agreed, and Smithy promised to bring his own patient to Baltimore for Blalock's first operation.

While the pair practiced the aortic valve procedure in Blalock's animal lab, two more surgeons announced success with mitral valve surgeries. In June 1948, Dwight E. Harken of Boston, a renowned World War II military surgeon, succeeded after the deaths of six of his first ten patients. Charles Bailey of Philadelphia announced success after losing four of his five patients along with operating privileges at the Hahnemann Hospital.[3] He took the lone survivor, a twenty-four-year-old woman, on a thousand-mile train trip to a medical meeting in Chicago, where chest surgeons erupted in thunderous applause.[4] Hardly a month after Bailey's mitral valve victory, Blalock was ready to try to repair an aortic valve.

*

The operation on Smithy's patient was scheduled in late July 1948, on the same day that Blalock was to install a second shunt into a boy with tetralogy of Fallot. The patient was one of Helen's earliest blue babies and had been operated on as an infant. She expected most shunts created by Blalock to last between five and ten years, maybe fifteen years. The boy was declining after only four years. He was among the first older children or adults to undergo a second operation. Helen was not present because she was beginning the drive to the Cape. But by the time she arrived in Cotuit two days later, she knew from a telephone conversation with her fellow, Engle, that both patients had died. In Blalock's retelling, "It was the worst day we have ever had."[5]

Helen's patient had died first. As soon as Blalock cut open the boy's chest on the same side where he had previously operated, he knew he had made a "foolish mistake," he wrote to Helen. He found the pulmonary artery thoroughly stuck to everything around it, so that it easily ripped. The boy quickly bled to death. The outcome confirmed the difficult road ahead for blue babies who outgrew their shunts and relapsed.

Smithy's aortic valve patient, scheduled in the afternoon, also died quickly. His heartbeat surged with the introduction of anesthesia. Then when Blalock's resident, Cooley, opened the chest, the heart's electrical system went haywire and the patient's lower chamber stopped pumping blood into the body (ventricle fibrillation).[6] Doctors could not revive him.

Blalock concluded that the aortic valve was not approachable without a new technique and refused to try the operation on Smithy, who died a few months later, in October 1948. Blalock did not operate on the more accessible mitral valve either until after Russell Brock, the top surgeon in Britain, demonstrated his own procedure in several operations during a 1949 visit to Baltimore.[7] (In preparation, Blalock sought Helen's opinion on the method and asked her to examine autopsy specimens with him.)[8] Hard working and thoughtful, Blalock had natural ability but was not as gifted technically as many of the doctors he attracted to his surgical training program.[9] Cooley, not Blalock, repaired a bulge in the subclavian artery to the upper body—an aneurysm—on one of Blalock's patients when he was away. (Blalock had seen it while removing an aortic obstruction in the patient two months earlier.)[10] Blalock's deputy, Bahnson, who stayed on after his residency to join the expanding full-time faculty, followed Cooley in making some of the first repairs to stretched and bloated walls in blood vessels. Blalock's legacy was evident now: to recognize and drive talented surgeons to unimagined heights.

In late 1948, after the failed aortic valve repair and as Blalock was beginning to realize the limits of his operative solution for transposed great arteries, he took Bahnson and a few others to Philadelphia to visit a friend who was working on an alternate technology. Beginning in the 1930s, inspired by the loss of a patient, John Gibbon with his wife, Mary, at the Jefferson Medical College had been building a machine he hoped could take over the work of the human heart during an operation. By the 1940s, with financial and engineering help from IBM, his machine could pump blood to the lungs for oxygen and on to the body, so that a surgeon could clamp shut arteries into the heart (stopping its activity) and, with a clear view of the heart, operate directly on it. It worked on cats.

In this era, doctors shared ideas but developed their own tools and processes. On this trip, Blalock glimpsed the future of heart surgery technique, but like many other surgeons, he came away skeptical that a machine to take over the heart's functions could ever be realized. Whether from doubt that it would work, his busy schedule overseeing an expanding surgical training program, or, as others have suggested, from a habit of ignoring ideas that were not his own, Blalock decided against pursuing a similar system himself. Instead he signaled approval for his deputy, Bahnson, to pursue one. Replicating Gibbon's machine was expensive and hugely time-consuming, and Bahnson in 1949 left for a stint in the Army.

*

As early as 1949 Helen asked Blalock if he was interested in ways to constrict or narrow the pulmonary artery. Some of her patients had large holes in their heart wall that sent too much blood into the lungs—the reverse of the problem facing blue babies. The lungs reacted by creating resistance, and this activity put stress on the right side of the heart. Extreme cases could increase pulmonary hypertension and lead to reversed blood flow. Blalock tried a procedure suggested by Helen's fellow Frank Dammann while Helen was away, but the patient was misdiagnosed and died.[11]

In summer 1951, Helen pushed again, telling Blalock she hoped he was working on this problem, because Dammann and Blalock's former chief resident, Muller, working together at UCLA, had now perfected a way to create a blockage in the pulmonary artery to reduce the blood to the lungs in a tiny baby. By making a wedge-sized cut to the pulmonary artery to reduce the size of the opening and covering the artery with a band of umbilical tape wrapped with polyethylene film stitched securely in place, they succeeded in reducing blood flow to the lungs by about one-third. Their so-called banding procedure would save many babies born with large ventricle-wall holes (ventricular septal defect) that at this time could not be repaired easily. Helen wanted this procedure for her patients too, since otherwise they would die.

This innovation marked the beginning of surgical breakthroughs involving the heart by surgeons outside Hopkins. It underscored the need for Helen to seek information and expertise on surgery at other medical centers to secure the best treatment for her patients.

Blalock's research focused on ways to normalize blood flow around the circulation system and fix the valves that regulated the passage of blood through the heart chambers and into the lungs and body. The procedures he tested and developed in the late 1940s and early 1950s were variations of Helen's idea to repair the plumbing of the beating heart; in addition to the path he created by linking two vessels (Blalock-Taussig shunt), he cleared obstructions in the aorta, opened the mitral valve, and created or patched openings that diverted blood to or from the lungs or caused it to reverse direction, stymieing heart-lung circulation. These operations were designed to relieve suffering and extend lives, but most did not address the cause or restore the heart to normal. Blalock was particularly gloomy about his treatment for children with transposed arteries. Although it was an important surgical advance and the first step in an operation to help such patients, it brought only moderate improvement. Moreover, most patients died. A true solution, one that would give children with this condition a chance at a long life, would be to disconnect their arteries and reattach them to the right places, he concluded. This was impossible without technological advances to "replace the pumping action of the heart and the oxygenating function of the lungs" during surgery. This was the machine he had seen in Gibbon's lab to take over the heart's functions, to oxygenate and pump blood to the body, so surgeons could repair the stilled heart.

To readers of *Life* magazine, Gibbon's machine was the stuff of science fiction. But a handful of doctors inspired by Gibbon, including John W. Kirklin, a young resident surgeon across the country at the Mayo Clinic in Rochester, Minnesota, were preparing for the day it would be a reality. In the early 1950s, Kirklin began filling notebooks with sketches of operations he could perform using it. Another surgeon in Minnesota, C. Walton Lillehei, at the University of Minnesota, spent what would become hundreds of hours over four years in the pathology lab to work out how to fix the heart defects detailed in Helen's book. He also assembled teams to diagnose and monitor patients and operate the human circulation machine he envisioned.[12]

When Helen in early 1950 sent Blalock correspondence she had with a commercial manufacturer about a pump that proposed to allow surgeons to stop the heart and operate on it, Blalock sent it back to her. He suggested she tell the company to discuss it with Bahnson when he returned from the Army in July since Bahnson had more interest in the artificial pump than anyone else in Blalock's department.[13]

*

While Blalock continued to operate around the beating heart, careful not to interfere with its electrical system (which controls the rate and rhythm of the pumping action of the heart so it can circulate blood and oxygen around the body), Gibbon reworked the process he hoped would allow surgeons to see and operate directly on the organ. By 1952, Gibbon was ready to try his circulation machine. Once it started to move blood to the lungs and into the body of his young patient, Gibbon temporarily tied off arteries to the heart, bringing it to a standstill, and stitched closed a hole in her upper heart chambers (atrial septal defect).

But of six or seven patients whom Gibbon operated on while using his machine, only one survived surgery, an eighteen-year-old woman on May 6, 1953.[14] The machine worked, but after losing two young children, Gibbon bowed out. High fatalities and an expensive machine that took years to build would normally discourage imitators. But the opposite happened: his surgery fired the imagination of surgeons who realized that direct repairs to the heart were possible if they could find a better machine to temporarily take over for the beating heart.[15] Kirklin asked Gibbon to share the plans for his artificial heart-lung pump.[16] For the next two years, he and his associates tinkered with Gibbon's model to build their own substitute for the human circulation system. Teams of cardiologists and surgeons, including Helen's former fellows Engle at Cornell and Dan G. McNamara, chief of pediatrics at Texas Children's Hospital, visited Kirklin at Mayo in Rochester and Lillehei in Minneapolis to observe these experiments.

Interest intensified when John Lewis at the University of Minnesota used an ancient method, hypothermia, to demonstrate operations on the open heart. Surrounding his little patient in ice, he reduced the body temperature to 79 degrees, thereby slowing the patient's breathing and his oxygen intake until the blood circulation neared a standstill. During this state of stillness, Lewis closed a hole in the wall of the heart in about six minutes. But hypothermia was not a viable method for most heart surgeries. The brain could not withstand a loss of oxygen longer than fifteen minutes.

On March 26, 1954, in the most significant if bizarre experiment of the era, Lillehei (also at the University of Minnesota) unveiled a method to take over the function of the heart that linked the circulation systems of two humans. This method, which Lillehei called cross-circulation, installed tubes between the patient and a volunteer (usually a parent) to permit blood to flow between the two bodies as if they were one circulation system. During the operation, the volunteer's heart pumped oxygenated blood into the patient's artery and her veins took away the used blood, which recirculated to the volunteer's lungs.[17] Lillehei used his method to repair a hole in the wall between the upper chambers of a young patient's heart; again, the operation lasted only a few minutes.

Bahnson and other Hopkins surgeons traveled to Minnesota several times to observe these experiments. On one trip, Bahnson observed first-hand the extraordinary risk associated with Lillehei's experiment: a tiny air bubble mistakenly pumped into the circulation along with anesthesia left a healthy donor suffering badly.[18] Upon his return to Baltimore, Bahnson worked on the idea of a human pump in the dog lab. He also had a primitive heart-lung pump machine in the making.

Lillehei's successes were stunning despite his questionable method, which risked the life of a healthy person to save another. He operated on forty-five patients in 1954–1955, and a majority (twenty-eight) survived, creating a sensation. Glowing accounts of the procedure appeared in national magazines and on television. From small holes in the upper heart, he moved to fix holes in the lower ventricles, the holes then handled by a banding procedure. Most spectacular, Lillehei began to fix blue babies. Using his human circulation method, he opened up the heart of several patients with tetralogy of Fallot and directly repaired their birth defects; he closed holes and opened the path for blood to flow to the lungs.

This was the promise of operating directly on the heart. It allowed surgeons to restore the heart to more nearly its natural state rather than craft makeshift repairs around it. If surgeons could repair defects in a patient with tetralogy of Fallot, Blalock's pioneering surgical techniques and the palliative Blalock-Taussig shunt could become outmoded.

*

It is easy to imagine Helen's alarm when in late 1954 her friend Andrus, the top adult cardiologist at Hopkins and, at the time, president of the American Heart Association, returned to Baltimore after hearing Lillehei describe his cross-circulation technique and confided that Blalock seemed to have given up the lead on open-heart operations.[19] Helen knew Bahnson was working on a heart pump and experimenting with cross-circulation in the hope of operating on the stilled heart. He was also deeply engaged in experiments on the beating heart: he was trying to repair the aortic valve that Blalock had refused to tackle following the death of Smithy's patient.[20]

Helen tried to goad Blalock into action. She asked him for a meeting to discuss Lillehei's method, offering multiple options including her home on a Saturday morning as Christmas approached. "My group is very anxious for a chance to discuss cross-circulation (the actual management of the families & the question of professional donors etc.), with you & Hank [Bahnson]," she wrote. "We'd appreciate a good talk before you leave on your cruise."[21] Helen's interest in an operative procedure that risked the life of a person other than the patient showed how little she knew about it. She would learn from Bahnson what Lillehei did not share publicly but which doctors whispered about: a donor, the mother of the patient, had suffered a stroke during one of his operations.

By spring, Lillehei had successfully repaired defects in ten children with tetralogy of Fallot. Blalock publicly congratulated him on the surgical repair, which he did not think possible given the position of the aorta. Helen hoped Lillehei would stop his new operation, no matter how much it helped patients, because of his method. But she remained open to alternative methods. Blalock sent several surgical residents to the new National Heart Institute to learn from a Hopkins-trained surgeon there named Andrew G. (Glenn) Morrow. With their help, Bahnson and a band of residents developed their own hypothermia method to enable open-heart surgery. They practiced operating on dogs with this method until they were convinced it was ready to be tried on patients, according to one of the inventors of this new process, James P. Isaacs. They still had to sell Helen on it, though.

One day in winter 1955, Helen and her staff of pediatric cardiologists trooped over to Blalock's animal laboratory to watch Bahnson and others operate on a dog. Helen was satisfied enough by the demonstration that, shortly afterward, she selected three patients for operations to repair holes in the upper heart (atrial septal defects). The method involved immersing patients in bathtubs of ice and water until their temperature registered between 29 and 31 degrees Celsius (84–87 degrees Fahrenheit). These procedures were the first open-heart surgeries at Hopkins and were performed by Bahnson, Jerome Kay, Isaacs, and Robert Gaertner.[22] Helen would have recommended patients who were critically ill if she followed her own advice.

Remarkably, even with their oxygen intake reduced by 50 percent during the operation, two of the three patients survived.

*

While Hopkins surgeons experimented with open-heart surgery, the all-too-familiar problems with anesthesia continued to bedevil their department. Surgeons continued to overrule anesthesiologists with Blalock's blessing despite protocols established by Proctor, head of the anesthesiology division within Blalock's surgery department, to reduce anesthesia-related deaths. Even nurse-anesthetists trained by Proctor in the new protocols were confused about whose direction to follow. In late 1954, Proctor argued with Blalock when a surgeon failed to administer a preoperative drug in an emergency, as Proctor and his anesthesiologist recommended, and Proctor told the surgeon the anesthesiologist was the better judge of drug. Blalock, his voice trembling with rage, warned Proctor never to advise the surgeon again.

A few months later, in March 1955, with Blalock out of town, Proctor complained to the hospital's director when Bahnson refused to follow anesthesiologists' directions during the trial of "an entirely new technique" for a dangerous heart operation.[23] According to Proctor, Bahnson insisted on having a nurse-anesthetist deliver drugs out of loyalty to Blalock despite agreeing that a physician would administer the drugs better and, indeed, asked Proctor to provide a certified anesthesiologist to oversee the nurse.[24] Proctor told the hospital director his division had no reason to exist if it lacked authority to decide who should deliver oxygen and drugs to a patient undergoing a dangerous operation. Blalock upon his return called in Proctor and told him he agreed the division was unnecessary. Proctor was out. By June 1955, when top universities were beginning to spend big money to build anesthesiology departments that employed dozens of certified doctors and conducted research to improve treatment, the handful of doctors on Proctor's team either resigned or were not reappointed.

As someone driven by methods and watchful for small steps that would advantage her patients, Helen sought information and advice from experts like Proctor, who was also a friend and neighbor, an expert on deafness, and once had nominated Helen and Blalock for a Nobel Prize. If Helen discussed anesthesia with Blalock, it is not in their correspondence. All too familiar with Blalock's fiery temper, Helen kept her own counsel in this battle, although her recommendations for treatment reflected the environment around her. But to physicians who consoled Proctor, the episode was chilling. For challenging Blalock on matters affecting patient safety, Proctor lost tenure, a full-time salary, and his lab. Shortly afterward, Blalock was elected chairman of the hospital medical board, making him the most influential person in the hospital.

While the crisis in anesthesiology played out and within weeks of the first hypothermia open-heart surgeries at Hopkins, Helen learned that an artificial pump to take over the heart's work was up and running at the Mayo Clinic. After more than two years of testing a heart-lung pumping machine he had adapted from the Gibbon model, on March 22, 1955, Kirklin began repairing holes in the lower heart walls of eight children. A string of successes followed difficult early operations, in which half of his patients died. Almost immediately a surgeon working under Lillihei in Minneapolis, Richard DeWall, devised a simpler version of Kirklin's machine.[25]

The advent of open-heart surgery was a game changer. Led by Lillehei, surgeons began to attempt to fully correct the defects associated with tetralogy of Fallot—a narrowed path to the lungs, holes in the ventricle wall, the position of the aorta—that caused the lower heart chamber to balloon and ruined its valves. Instead of complicated moves on the outside of the heart to improve blood flow, surgeons could repair holes and compromised valves inside the heart itself. The implication for adults with diseased hearts was staggering.

Helen watched as patients flocked to the Mayo Clinic for open-heart surgery just as they had come to Hopkins only a decade earlier for blue baby operations. They also flew to Minneapolis, where Lillehei dropped cross-circulation in favor of DeWall's simpler heart-lung machine. The Cleveland Clinic was next to offer open-heart operations. By the end of 1956 at Baylor University in Houston, Blalock's protégé, Cooley, operated on ninety-four patients using his own adaptation of DeWall's heart-lung machine.[26] Helen's own institution, meanwhile, struggled with imperfect pumps.

*

From the beginning of their relationship, Helen had believed and tried to convince Blalock that with her book and his surgical skill, they would offer the best patient care. None of the surgeons who pioneered the open-heart procedure had access to Helen's diagnostic skills. Gibbon had stopped operating partly because he misdiagnosed a patient who died during the procedure. Mayo Clinic, too, lost patients because of incorrect diagnoses or understanding of how the malformed heart worked.[27] Blalock had let their very efficient partnership deteriorate. Privately Helen was no doubt dismayed, but she had patients and an image to protect.

On behalf of those patients Helen now scrambled to learn about open-heart surgery methods from afar, through pediatric cardiologists she had trained, including Engle in New York and McNamara in Houston. Blalock too scrambled. He was not happy about others taking the lead in this new field, according to Vivien Thomas. As it became clear that surgery on the open heart would become the norm, Blalock rallied resources and assigned a second surgeon, Frank Spencer, to help Bahnson develop a heart-lung pump.

An energetic man, Bahnson operated on some of the first blue babies himself and carried a heavy surgery load under Blalock, in additional to temporarily leading the catheterization lab and developing the heart-lung pump. He had demonstrated the Blalock-Taussig shunt with Blalock in London and Paris and performed the first operation in France to close an open fetal duct when an exhausted Blalock left him in charge and went for a stroll in the Tuileries Garden.[28] Bahnson practiced repeatedly in the lab before performing complicated new surgery. He was one of the most experienced heart surgeons anywhere. Helen admired him, but the first open-heart surgeries using a heart-lung pump had such a high mortality rate that she grew skeptical. A Gibbon-style pump developed by Gaertner and others and used by Bahnson and Spencer in 1956 resulted in many deaths, according to Isaacs, and caused "consternation" among surgeons, Helen and her staff, and nurses.[29] Helen quickly realized that some patients with holes in their upper wall chambers would still be alive if they had been operated on using hypothermia or a closed-heart procedure, as before, instead of the new heart-lung pump. She continued to refer patients for open-heart surgery using hypothermia, but she remained cautious about the heart-lung pump even after Bahnson transitioned to a smaller, refined machine adapted from the Mayo version and used successfully by Cooley and McNamara in Texas.

"Bahnson was ready to go long before she ever turned a patient loose," Blalock's lab assistant Vivien Thomas recalled.[30] "She did hold back a long time before she ever let them pump anybody around here." She also concluded that infants as small as the first blue baby, Eileen Saxon, were unlikely to survive open-heart surgery regardless of the method.

*

As she considered open-heart surgery for her patients, in spring 1957, Helen asked former fellows to return to Baltimore for a special meeting on the heart, an informal discussion for them to present what they had learned about open-heart surgery using a pump and hypothermia. She invited them to bring the surgeons they worked with too, and she opened the gathering to pediatric cardiologists who had not trained under her but who developed top programs and had visited her clinic, among them John D. Keith at the Hospital for Sick Children in Toronto and Alexander Nadas at Boston Children's Hospital. Everyone wanted to know what worked and what didn't without waiting for it to show up in medical journals; she and Charlotte also would preview research on the development of lungs in infants with birth defects. Blalock agreed to come. Helen's timing was prescient: the information she collected in these years would guide her own practice and lead to another major pronouncement on when to recommend patients for surgery.

It was in this challenging patient care environment that Helen considered how to answer a friend who sought her advice about where to obtain an open-heart operation for a young boy. The friend was noted British pediatrician Alfred White Franklin, who had met Helen in 1934 while researching at Hopkins. His patient needed an operation to correct a hole in the wall of his lower heart chambers (ventricle septal defect). This was the most performed procedure using the open-heart method at the time but an extremely difficult one. After reviewing his case, Helen advised Franklin that his patient's best chance for success was at the Mayo Clinic. News of Helen's choice rippled through hospital hallways. Helen's fellows knew why she did it: surgeons at Mayo had more experience and a vetted machine, and the link between experience and higher success rates was indisputable.

Coincident with her decision, Helen was investigating problems with operations to close the hole in the ventricle wall and was preparing to issue guidelines about which patients should be recommended. This difficult operation had a significant mortality rate due to heart block, a sudden and usually fatal stop in the heart's electrical system that sometimes followed the surgeon's incision.

Helen's recommendation to Franklin underscored that her loyalty lay with her patients. Top medical centers competed for a national flow of patients, and hospitals were beginning to earn money from insurance companies for health services. Helen's colleagues would have been upset not only about their reputation but also the potential loss of revenue. Unlike other times when Helen delayed or rejected surgery on her patients, however, surgeons dared not challenge her. Bahnson himself gave not a hint of annoyance and "continued to be very nice to Helen," Charlotte recalled. But as his friend, Charlotte knew that being passed over by the top pediatric cardiologist in the world was a blow. "He was wounded," she said.

The staff at the Mayo Clinic was reportedly not any happier. This was because after Helen advised the doctor and his patient to go there for open-heart surgery, she went with them, all the way into the operating room. This was Helen's opportunity to learn firsthand about open-heart surgery from the experts. Everywhere the boy and his family went in Rochester, Helen led the way, asking questions, assessing the state of the art, and overseeing the child's care. The Mayo staff was unprepared for Helen's intensity and thoroughness, attributes those in Baltimore had long factored into their day. Pediatric cardiology fellows at the Mayo Clinic were overwrought trying to meet Helen's expectations and called doctors at her Baltimore clinic for advice. While attending to her patient, Helen wandered the hallways politely but boldly examining every aspect of their process, facilities, and patient care. According to Charlotte, who communicated with Mayo staff during this time, "she drove them crazy."

20 DOUSING THE FIRE

The potential to help patients using the open-heart method was enormous. Sewing shut a hole in the wall between the heart chambers to curtail excess blood flow to the lungs was easy compared with old ways of patching holes or banding arteries. Because repairs were easier, Helen feared infants would be targeted for surgery, although most could not survive any repair and, as she had learned, some might never need one. Little holes healed themselves, and little bodies adjusted.

A decade of experience with blue babies and children with similar conditions also showed Helen the complexities and dangers of interfering with blood flow to the lungs. In Brussels in 1958, she and Charlotte, using lung samples from Toronto and Baltimore, unveiled a study of one hundred patients who died after having had shunts installed. Some died within a month due to injuries caused by a sudden increase in blood, and others died a decade later after their underdeveloped lungs grew increasingly resistant to the extra blood coming through the shunt and they experienced the same difficulty breathing and blue skin as before surgery.[1] With this in mind, and as Helen learned about the impact on the lungs of newer operations to correct or repair heart defects, she again advocated a wait-and-see approach to surgery on children.

In an urgent 1958 editorial in *Circulation*, the journal of the American Heart Association, and at medical meetings she advised pediatricians that the single most important question to ask when recommending an open-heart operation was which surgical method was safest. It does not appear she publicized data identified with her own hospital. But pressure from

Helen was surely one reason Hopkins used hypothermia well into the 1960s to repair atrial wall holes in children.[2] In the late 1950s, doctors elsewhere too were discovering that the mortality rate for some simple procedures increased when the artificial pump was used, possibly because of flaws in the heart-lung pumping machine application but also because of the age of the patient. In response, some surgeons advocated hypothermia for simple open-heart procedures.[3] Some even reverted to procedures on the closed heart. At the University of Pennsylvania, the death rate for an uncomplicated mitral valve stenosis on the closed heart was 3.4 percent compared with 12.4 percent using open-heart surgery.[4]

Helen was haunted by the words of the eminent British cardiologist Sir John Parkinson who, upon greeting her in London not long after blue baby surgery began, told Helen that her idea had "lighted a match and started a fire."[5] As on the long-ago evening in Cotuit when from her window, Helen saw a spark on the pine path through the woods, she worried that the fire would get out of control.

*

Her first warning against unnecessary surgery had been directed toward pediatricians. Now she appealed to all heart doctors and to the public. This time, because of problems with heart-lung machines and a growing body of information about the unanticipated impact of some new repairs on the heart, Helen argued ever more forcefully *against* operations for patients who were only slightly incapacitated. In *Circulation*, Helen specified the minimum function a patient could have and still be better off without an operation. For example, she said a blue baby's blood oxygen saturation could drop as low as 66 percent before an operation was warranted. She also specified the acceptable rate of enlargement of the heart and the amount of pressure on the lungs. These two problems could subside on their own under certain conditions.[6] To newspaper reporters waiting for Helen after she delivered her paper, she gave a shortened version and addressed her concerns to parents. Headlines followed.

Health records are private, but some patients chose to share their experiences publicly, and from these, we know that Helen's decisions meant some children suffered for years rather than undergo risky operations. Elmo Zumwalt III, the son of US Navy Admiral Elmo Zumwalt Jr., one of his generation's most distinguished military leaders, experienced frequent bronchial attacks and lived with a heart that ballooned to twice the size of normal for six years before surgery. His problem was that a small hole in the wall between the upper chambers (atrial septal defect) allowed extra blood to drip into his lungs. This forced the right side of his heart to work harder. Helen believed his inflamed heart would revert to normal either when the hole closed on its own or after appropriately timed surgery.[7] She grew concerned and suggested the family consider the operation in 1958 when the boy turned twelve, perhaps because she detected a change in the rate of enlargement of the heart and the amount of pressure on the lungs.

As Helen advised others to do in her *Circulation* essay, she next considered the method. Since more children survived the hypothermia method than the heart-lung pump for this operation, she and the surgeon, Frank Spencer, chose hypothermia over the new pump in use at Hopkins. Spencer put the boy on ice for less than ten minutes to slow his circulatory system enough to allow him to operate on the heart and still permit adequate oxygen to flow through his body.

This family's agonizing two-hour wait outside the operating room with three other couples whose children were also undergoing open-heart procedures revealed the risk of these operations. The Zumwalts watched one set of parents gather their belongings and leave in tears after a doctor emerged to deliver bad news. A second couple got uncertain news. Only two of four operations that day were successful. Zumwalt Jr. knew by the smile on Spencer's face as he approached that one had been his son.[8]

The risk of surgery was so great that even for very seriously ill patients, such as those with an aortic valve blockage or a high opening in the heart wall separating the ventricles, Helen suggested doctors first limit their patients' activity before resorting to an operation.[9] This was extremely conservative advice. She was not only concerned about the technical quality of surgery, which she knew would improve. She was worried about the patient's long-term survival.

Her editorial in *Circulation* hinted at details to come. She was working on a paper that would urge severe limits on the most frequently performed operation using the heart-lung pump: closing a hole in the ventricle wall. For now, she warned that each operation should deliver only the right amount of blood to the lungs. For each patient, that amount differed.

*

Around this time, in 1958, Helen had lost several children who had gone into heart failure after operations that resulted in too much blood to their lungs. She paid special attention to how much blood her surgery patients could tolerate and instructed her fellows to do the same. Among the most difficult cases she handled were second operations for grown-up blue babies—older children with tetralogy of Fallot. As she and Blalock predicted, some children who had the Blalock-Taussig shunt installed at age five or six began to outgrow them in their teens. One day in 1958, Helen brought a fourteen-year-old patient to see Blalock and asked him to operate. The boy had had a shunt installed as a toddler. He returned to Helen's clinic short of breath and exhibiting the same blue tint she had seen prior to his operation.

Helen had ruled out what doctors called a "total correction" of tetralogy of Fallot, the open-heart method to repair a valve, widen the narrow pulmonary artery to the lung, and close the hole in the heart wall. This was the best method for young children without previous surgery, but it was risky for a teenager who had lived with a shunt for years. The option for teens or adults whose childhood shunts were failing was to create a second shunt by sewing together the body's two main blood vessels. Usually surgeons sewed a section of aorta that turned down into the body to the left pulmonary artery. They could also attach the right pulmonary artery directly to the aorta (Waterston shunt).

Either was more difficult than the Blalock-Taussig shunt because these major vessels were larger and blood flow between them could be unlimited. The surgeon's cut into the artery largely determined the amount of blood going to the lungs. Too much blood and the child might die of heart failure. Too little and the operation would be futile. Second operations on blue babies were also dangerous because the surgeon had to get around scar tissue from the initial procedure and confront unknown paths that collateral blood vessels had taken in the interim years. Blalock hated these cases. But he agreed to operate on Helen's patient and sent his chief resident that year, J. Alex Haller Jr., to confer with one of Helen's fellows.[10] Haller prided himself on his ability to get along with Helen's female fellows. After reading Helen's book, he had elected to rotate through her pediatric clinic. He was training to be a children's surgeon.

The amount of detail reviewed in preoperation meetings could be numbing. The resident surgeon and Helen's fellows typically discussed the patient's blood and oxygen status and the size of his vessels, and therefore, his or her capacity for blood flowing to the lungs. At the meeting over this boy's operation, they discussed the complexity of attaching a new artery amid scar tissue from the first operation. Another complication was the boy's size; he was very muscular. If Helen or her staff had recommendations on how much blood should flow into the lungs or how big the shunt should be, this was the time to bring them up. According to Haller, Helen's fellow, whom he declined to name, said nothing about this.

The operation was extraordinary. Standing beside Blalock that day, Haller saw the patient bleed excessively, and his blood was almost black, a sign of poor oxygen flow. Blalock was not happy. The doctors managed to sew the two blood vessels together, but not before some blood spilled into the lungs. Finally, the bleeding stopped. The second shunt appeared to be taking. The boy's color was not yet normal, but the small change in the lips and tips of the ears left Blalock satisfied.

Worn out, Blalock instructed his chief resident to put in the chest tubes and returned to his office. Haller drained the fluid and began closing the patient. The muscles of the chest wall came first. Then he pulled the ribs together. Haller was about to close the skin with stitches when he looked up to see Helen's fellow encased in a gown, hat, and mask poking her head through a slit in the door of the operating room. To his amazement, she asked him how the operation was progressing.

This type of interference was unheard of: doctors did not work in teams the way they do today. No doctor who put a patient in the hands of a surgeon would send an agent around, seeking details, before the patient was out of the operating room.

Haller reassured her; despite extensive bleeding, the shunt had held. They had been fortunate to come this far. He began stitching up the boy. The anesthesiologist also agreed the boy was handling the shunt well.

She persisted, "How large is the shunt?"

Haller couldn't tell how big it was; he was only happy the boy appeared significantly improved, indicating the shunt was working. But he took a guess. He told her it was 4 millimeters wide.

"Oh," replied the fellow in a startled tone. "Dr. Taussig thinks that is too large. She thinks it should be no larger than 3 millimeters." Anything larger, she said, and the teen would suffer a heart attack from too much blood flowing into the heart chamber.

"Too late now," Haller responded as he returned to his sewing.

The fellow disappeared. Haller had twenty minutes of stitching remaining in the four-hour operation when he pushed aside the drapes around the operating theater for some air, and out of the corner of his eye he saw Blalock in the adjoining room donning a gown and mask. The surgeon pushed open the door with his back and began scrubbing.

"Open up, Alex," he called in an unhappy tone. "We have to go in again and decrease the size of the shunt." Haller was dumbfounded. Other doctors too appeared shocked. But the patient was stable and in fifteen minutes, while Blalock scrubbed, Haller took out the stitches and reopened his chest.

"Now Alex, where is the shunt?" Blalock asked him.

It was hard to see, but Haller felt it, putting his finger over it.

"Fine," Blalock responded. "Give me the suture."

Haller watched as Blalock sutured in the area of the shunt. To Haller, it looked as if the surgeon took one stitch. Maybe two.

"That's it," Blalock said.

As far as Haller could tell, it didn't change anything, but Blalock could say he tried to fix it.

An hour and a half later than expected, after stitching up the patient a second time, Haller left the operating room, put on a clean white coat, and walked fifty feet down the hall to Blalock's office. As he entered the secretary's suite, he noticed smoke billowing from Blalock's inner office. There was so much smoke he worried briefly that the building was on fire.

"What in the world is going on?" Haller asked as he pushed open the door.

Blalock was sitting in a chair behind his big desk. With the cigarette moving up and down with his hands, Blalock recounted how Helen had come to his office after he had left the operating room to say that the shunt was too large and, in her experience, it would not work. She asked him to fix it. She was adamant. She had seen the boy's lungs.

Blalock considered the shunt fine. Mostly surgeons worried about the shunt being too narrow, not too large. It would have been hard to predict how big the shunt should be without knowing the size of the boy's vessels. A 4 millimeter opening was big, but so was the boy. It was a judgment call.

In this case, the surgeons had no way to fully measure the flow under the scarred tissues. They had done the best they could.

If the boy died, Blalock would be responsible.

Helen was usually right. Blalock knew she would not give up. Yes, he told her; he would do that for her.

"Alex," Blalock told his resident as smoke trailed above him, and he gestured with his cigarette. "If I get to heaven, it was because I could live with Helen Taussig."[11]

*

Around now, Blalock's wife was ailing. Helen sent him notes, wishing her well. Mary Blalock died in late 1958, and Blalock became depressed. Earlier that year, at a day of horse racing and festive parties in Baltimore County known as the Hunt Cup, Helen's fellow Sissman observed Blalock, alone and formally dressed, unfold a stool, sit down, and take a silver flask from his coat pocket. Blalock had always enjoyed a drink in his study after dinner. But now Vivien Thomas, who had beaten his own drinking problem, drove to his house to confront him about it.[12] While his deputies performed daring open-heart surgeries, Blalock returned to the laboratory. He spoke of retiring.

Helen's next target in her battle against unnecessary or aggressively timed surgery was the most frequently performed open-heart operation: the surgery Helen had observed at the Mayo Clinic, closing a hole in the bottom heart wall (ventricle septal defect). A child with a hole in his bottom heart wall had too much blood flow to his lungs. Closing it could save some children from developing pulmonary vascular disease and dying young of heart failure. But small children could not withstand an open-heart operation, and all of them faced the risk that the heart's electrical system would backfire. The surgeon could unknowingly interfere with the electrical conduction system when he stitched, blocking the electrical impulse and causing the heart to slow or stop. Children died of heart block.

Many children did not need the operation. Others could wait. In 1959, in a third guide for pediatric cardiologists, Helen, with Catherine Neill as lead author, spelled out when to recommend this operation. The pair based their conclusions on Helen's experience but also on studies of early operations at Mayo Clinic and those by McNamara and Cooley in Texas.[13] An operation was obviously indicated for children with a hole so big that the two pumping chambers in the bottom of the heart merged into one. But many small holes were entirely manageable without surgery. In fact, surgery could interfere with the body's natural response. Helen knew as early as 1955 that children born with small holes experienced delayed lung development. Her former senior fellow, Dammann, assisted by Charlotte (and, separately, Jesse Edwards at the Mayo Clinic), had discovered that the lungs of infants born with holes in the walls of their hearts do not develop normally. Instead, their ability to exchange gas and allow oxygen into the blood developed only gradually, sometimes over years; meanwhile, their thickened walls defended the lungs from the extra blood coming from the hole in the heart wall. This was nature's way of shielding the infant's immature lungs from being overwhelmed. As long as the lung beds eventually returned to their natural state, children with small holes did not need surgery to reduce blood flow to the lungs. They would grow to handle it.

That left a variety of cases in the middle: children with different sized holes who remained stable in childhood. For this group, Helen urged pediatric cardiologists to wait as long as possible before recommending surgery. At every visit, doctors had to weigh the risk of an operation against the risk of developing hypertension in the pulmonary artery, an incurable condition. At the time of her warning, one of Helen's patients, here named Stuart Reedy, had already lived one decade with a small ventricle hole she diagnosed when he was six weeks old. At her suggestion, the boy's parents allowed him to lead a normal life and kept close watch on his health. So far, he was thriving.

*

Blalock's surgeons, too, were writing papers on what worked and why, and trying to solve heart block. Some regarded Helen's advice as extreme and potentially dangerous for the middle group of children. To combat her, these surgeons needed to cross the deep river that separated them from Helen's pediatric cardiologists. Partisanship (and honor) had become so ingrained at Hopkins in the 1950s that the election of a president of Alpha Omega Alpha, the medical honor society, depended on which group, medical (including pediatricians) or surgical, had the most members. With Helen's influence in defining her field and new ways of work in clinical practice, these camps had solidified and spread. Helen's fellows took her conservative attitude with them as they fanned out to work with surgeons at Yale, Cornell, Baylor, and UCLA, among other top US universities, and at universities and hospitals abroad.

The resistance from Helen's disciples was so impenetrable that four years after she issued her guidelines, Cooley and Spencer convened a roundtable discussion at a professional conference in Chicago to try to win them over. Cooley and other surgeons were convinced they could and should operate sooner on stable young children with holes in their ventricles because newer techniques reduced the chance of heart block. They were *not* operating on this group, though, because Helen's former fellows were "hiding" asymptomatic children.

Spencer called it delusional to think that a hole in the bottom of the heart wall was not a serious condition requiring urgent attention, and he rejected Helen's idea that there was a detectable pattern of when ventricle wall holes might lead to pulmonary hypertension and irreversible damage. (Helen maintained it was between four and eight years.) To him, there was no reason to wait. He said one of three children advised by a pediatrician to wait more than five years had died before surgery.[14] "If you have a hole there, I don't think you have to wait until heart failure starts to operate," he said. Cooley argued that these operations were safer than closing the open fetal duct, an operation routinely performed, "so I think we can stop worrying so much about whether we are maiming these little patients." He boasted of no fatalities in eighty-five operations to close heart walls in young children.

But Dammann, then at the University of Virginia, disagreed that the operation was safer than closing fetal ducts. Under his questioning, a surgeon who had passed through Helen's clinic, William Glenn of Yale, estimated mortality rates of children under the age of seven with no symptoms at 5 percent. He and his pediatric cardiologist partner, Whittemore, Helen's first fellow, limited operations to children over the age of eight. Dammann too refused to tolerate a 5 percent chance of death for a young child experiencing no problems. He would not recommend operations for this group until the mortality rate dropped to 1 percent.

*

Elmo Zumwalt III died of cancer related to a chemical used in combat, Agent Orange. Stuart Reedy was twenty-one when surgeons finally closed the hole in his heart. He became a scientist.

The model developed by Helen for collaboration with surgeons and other specialists would become the gold standard. Children with heart defects would be examined not only by a surgeon who knew how to fix them, but also by an expert in children's hearts who could consider the individual's long-term outlook, the ailment, and the surgeon's experience with the procedure.[15] As surgery expanded, doctors in other specialties adopted similar models. Teams became the norm. But in the early days, it took a strong will to create and defend a line between patients and the operating room. Helen's conservative attitude and her influence infuriated surgeons who believed more patients could benefit from surgery.

Cooley, the first intermediary appointed by Blalock to "handle" Helen during blue baby operations, regarded her as too intrusive and close to her patients. One of the most aggressive and best surgeons of his generation, and the first to transplant the heart of an infant, Cooley befriended and admired Helen even while he regularly challenged her, including on whether to operate on tiny babies. He called her "a tough nut."[16] Others who disagreed with Helen blamed her decision-making on an overly close relationship with her patients. They described her as possessive and motherly.

Helen was protective of her patients. But her mothering could be cold and calculating. Her judgment was the result of a method that reflected her precision, her disdain for invasive or unnatural solutions, and her experience observing the path of diseased and defective hearts over decades. Surgeons too were concerned with outcomes, but they did not share Helen's sensibilities and lacked her long-term database, one she consulted from memory. As a scientist, she continued to pose questions and evaluate data beyond the immediate clinical outcomes that surgeons pursued. She was building patient trust as well as safe and effective treatments for her patients. This explains why someone who helped start heart surgery worked so hard to limit it. She had the long term in mind, the bond of trust between physician and patient. Like the daffodils she and her fellows planted on her sloping lawn, she wanted heart surgery to take root and benefit others long after she was gone.

"V.S.D. and T/F are doing better," she wrote Charlotte in November 1959, using shorthand for holes in the ventricle walls and tetralogy of Fallot, "but we are not free from complications and have to re-open a great many, too many chests—we get some brilliant results but I still think we are over-enthusiastic and forget the disasters."[17]

Heart surgery on young children would become routine. But in the beginning, when tools were primitive and the impact on the body revealed only after years of trial and error, Helen's observations allowed her to speak with authority and instilled hope, an attitude Helen believed was a condition for success. Some patients believed operations succeeded as often as they did *because* of Helen's watchful eye.

"Your presence gave me courage and confidence," a lawyer operated on as a child wrote to Helen after his second surgery decades later. "I shall long remember how you walked with me as I was taken to each surgery and how you stayed with me during all the preparation and how you were there in intensive care when I woke up," he told Helen. "I deeply believe that all went as well as it did due to your presence and due to your subtle supervision of my surgery and recovery."[18]

21 GIFTS FROM THE HEART

The cards began arriving in Helen's mailbox in mid-December. She opened them on her sofa in front of the brightly lit fireplace or in the straight-backed chair on the left that she preferred for her posture. She hung them like ornaments on string nailed around the windows, covering sills and sashes, and the books in tall shelves on either side of her desk. Eventually the string of cards extended into the living room. By Christmas Eve, when she left to spend the holiday with her sisters in Boston, the room was aglow with red and green scenes, glitter, snow, Santa, children on sleds, and photos of happy faces. Glancing up, Helen could see how her patients had grown. Evergreens from her garden filled a crystal bowl on the table. It was the scent of joy.

"That's my family," Helen told her niece, Polly, as she inspected the hundreds of cards.

Initially parents wrote the cards. One mother boasted that her son had returned from a six-mile hike no more tired than the other boys. Another that her son worked in a drugstore even after his stroke. When they were old enough, the children penned annual letters, at first to introduce themselves. "My heart is as good as new and all thanks to you," one child penciled on a hand-made Valentine. "I'm just one of the gang," wrote a little boy pictured on a field with three friends, squinting like Spanky of the TV show, *The Little Rascals*. Easter, when Christians celebrate the miracle of the Resurrection of Christ, was another popular time to remember Helen. "Thank you for giving me the chance to live," one girl wrote. A neighbor one Easter dawn described "the results of your miracle": the once-listless little boy who lived upstairs was riding his bike, fishing, and hiking. On Mother's Day, too, they honored her. Helen bought three wintergreen barberries—*Berberus julianae*—with the twenty dollars a parent slipped into a card. In gratitude, on their children's birthdays, mothers wrote checks to a "mercy" fund Helen maintained for needy patients. "It is now 30 years since [. . .]'s surgery," one wrote.

Any given day Helen opened her mail to learn which college a former patient would attend, the countries where they traveled, or the next stage of their professional successes. An artist sent Helen a commercial greeting card she had designed. Another painted Helen a miniature mountain scene. "I will never, ever forget you," the young woman wrote. Inside their letters, children tucked photos of themselves: girls in sailor suits, bows, and pearls, carrying dolls; boys in scout uniforms and cowboy hats, on a slide or playing drums. They were pictured in gardens, sitting on a pony, in swimsuits, canoes, on a motorcycle, emerging from a church, reading a prayer book, on the boardwalk, and on their wedding day.

Helen in response congratulated former patients when they finished college, recommended them for jobs, advised on whether to have children, opined on how much housework they could handle (all of it), and shared her weight-loss regimen. She encouraged them to play tennis, golf, hunt, throw discs, dance, and swim, to their limit. She sent practical wedding gifts, like a stainless-steel butter dish. The news that one former patient was to be married "brought me great joy," Helen wrote her. The young woman's letter was filled with details a relative would welcome, a description of her future husband, their new apartment.

Helen's relationships went far beyond the doctor-patient relationships that defined the era. She traveled by taxi and plane to see them, she took calls from teenage girls late at night, and not infrequently she advised parents to let their children grow up. One cold, snowy day in 1960, she invited the parents of a newborn infant into her home to explain his condition, transposed great vessels, and recommend emergency surgery.[1] She hosted patients and their families for dinner in Baltimore and was a guest for dinner in their homes. She invited some to Cotuit, even insisting the British patient operated on at the Mayo Clinic and his mother recover there for months. In photos Helen stands next to a grown-up patient and her dogs on the beach. When a former blue baby named her newborn for Helen, Helen flew overseas to attend the christening. She counseled parents against treating their children like china dolls instead of allowing them to lead normal lives.

She helped find work for patients whose conditions were inoperable or their surgery only partially successful. Despite her trouble breathing, a woman who sought work with handicapped children had much to offer, Helen told the New York City Department of Education. "I hope you will give her application for teaching special consideration," she wrote. She identified doctors for faraway patients and convinced them not to charge those who could not pay. Helen advised patients on whether to have a second operation in the hope of expanding their oxygen supply. Helen was seventy-seven when she wrote a point-by-point assessment of one man's chances for surgery—which she judged good.

Some children with heart malformations suffered from developmental disabilities, leading to questions from doctors about whether the operation was justified, since children with diminished intellect might never live independently. Helen defended her patients, refuting the mistaken notion that children with tetralogy of Fallot or related birth defects also suffered intellectually because of a shortage of oxygen in their childhoods. There was no correlation, although a minority of children with Down syndrome did suffer heart problems. She set criteria for deciding when children with disabilities could benefit from operations with an eye to ending suffering but also to ensure the patient could lead a productive life, leaving the decision to the patient and parents.[2] This was the aim of all surgery, she believed.

One of the most common questions from women who had undergone surgery for tetralogy of Fallot was whether they were healthy enough to have children. Helen's answer depended on age and whether the patient had the original shunt or a full correction during open-heart surgery. "I gave them my blessing," Helen wrote in the file of a patient who with her fiancé discussed the question with Helen. Birth announcements followed. "Thanks . . . I had a successful pregnancy," one woman wrote.

Helen reassured former patients, male and female, there was only one chance in twenty their own children might be born with a defect (later estimates were as high as 10 percent). Consistent with her naturalist approach, she recommended a diaphragm instead of a contraceptive pill to women patients (or their doctors) who sought her counsel on sexual activity and pregnancy. She considered abortion less risky than long-term use of the pill (first approved by the Food and Drug Administration in 1960).

Helen's patients felt themselves in a special club, whose members lived in gratitude and believed that they had given something back by contributing to Helen's grand experiment. Some remained in touch from 1938 when Helen reexamined patients in her clinic for her book. For decades, they participated in her long-term studies to assess surgical outcomes by type of operation and other variables, information that helped surgeons and cardiologists improve care. They served as drivers when she alighted in their towns, allowing Helen to visually assess their well-being against what they wrote in annual surveys.

The club included rheumatic fever patients who had withstood Helen's strict regimens. The little boy who had been frightened by Helen when she assigned him four years of off-and-on bed rest had managed one of Baltimore's biggest companies for more than a quarter century when he wrote to her in the 1970s. Raymond Gebhardt had been thrilled to watch his sons play in ways he never could. He was now a grandfather.[3]

*

In the early years of Helen's relationship with Barbara Rosenthal, the first surviving blue baby, the pair lobbied against restrictions on animal experiments, appeared as featured guests of a celebratory cardiology conference in Buffalo, and discussed the new operation on television.

In 1947 Helen had a long talk with Mrs. Rosenthal after she reported her daughter was fearful and hesitant and recommended that Barbara see a psychiatrist. (Difficulty adjusting to normal activities was not uncommon among former blue babies.) When the family moved in 1948 to Los Angeles, Helen asked a friend, cardiologist Louis E. Martin, to treat Barbara. From afar she investigated Mrs. Rosenthal's reports of dizzy spells and reviewed the teen's X-rays (normal). She also nixed Mrs. Rosenthal's suggestion for plastic surgery to reduce the size of Barbara's lips, telling her she and Blalock thought it too risky. Their slightly extra fullness may have been from oxygenated blood flooding a space initially deprived of it. Barbara pronounced herself fine in an interview with *Newsweek*. Ever since her operation, she said, "I have been going like crazy."[4]

With her high energy, soft voice, and ready smile, Barbara, or "Barbs," as her family called her, was the life of her household, indulging her shy brother's interest in astronomy and insisting he attend school proms. Other than his sister's occasional short temper, he remembered no lasting issues.

Whatever anxieties she may have had seemed to have been overcome in high school, where she imagined herself a scientist and studied chemistry and physics. "Congratulations!" Helen told Barbara when she graduated. "Love and best wishes," she wrote. At the University of California, Berkeley—Helen's alma mater—Barbara told no one about her heart surgery. Curious about the world, she pursued a broad education and practical understanding of political and economic institutions, certain it would help her solve its puzzles, and she joined the staff of the *Daily Californian.*

Her parents worried over whether she could manage the hills and climb the stadium stairs. She could. (Helen only advised against basketball.) Like other parents of blue babies, the Rosenthals also sought reassurance about Barbara's prospects for marriage and children. Helen convinced them it would happen. "I certainly hope that she will marry and have that fullness of life, and I see no contraindication to that," she wrote to Shirley Rosenthal in 1952 when Barbara was a sophomore. As to children, Helen would examine Barbara when the time came.

Helen and Blalock both congratulated Barbara when, at age twenty-three, she married her college sweetheart. Healthy, happy, and busy, Barbara wanted more. She wanted a baby. But her father was adamant that Barbara not become pregnant, for fear she could not survive the pregnancy. According to Paul, her brother, her subsequent pregnancy greatly upset their father, and after consulting doctors who advised it could be harmful to her health, Barbara had an abortion.

Whether Helen was consulted is unknown, but in Los Angeles for a medical conference in 1959, she commandeered a ride from the airport from the Rosenthals and promised to visit Barbara at her apartment. Barbara looked "well & happy & pretty," Helen reported. "If she ever raises the question, tell her I'm sure her lips will correct themselves," Helen told her mother. "Truly all her looks need is more radiance underneath & that is coming fast."

While working in public relations for Cedars of Lebanon Hospital in Los Angeles in 1960, Barbara learned at a heart symposium about the full correction of her condition made possible with heart-lung machines. She also learned that children who had had early surgery could not expect normal life spans and realized the conferees were speaking about her. Although healthy, Barbara began to seriously consider a second operation to correct her tetralogy of Fallot. Her husband, parents, doctor, and surgeon approved it. Helen advised against it, though her reasons for doing so have been lost. She would have remembered Barbara's unusual and substantial collateral circulation that developed during childhood to compensate for the lack of a passage to her lungs. Hank Bahnson, the surgeon who frequently operated on Helen's patients, would soon publish results of five years of second operations and invited Helen to add her name to the paper. It showed in part that patients with pulmonary hypertension usually did not survive a second operation.[5]

Helen's assessment upset Barbara, but she insisted on the operation. Barbara was drawn to the promise of a normal life, without restriction, to do things she had been unable to do. Having a baby was chief among them.

On the Friday night before the surgery, Barbara's parents prayed at their synagogue and came away hopeful. But when doctors turned off the heart-lung machine following a nine-hour surgery, Barbara's heart was unable to resume its normal work. Blood rushed into the right ventricle that pushes blood to the lungs, overwhelming it. The chamber had been weakened long ago by inactivity and, after the first operation, by the extra work it undertook to get past blockages (pulmonary hypertension). There was a clot in the path to her lungs and hypertensive pulmonary vascular changes that made it unlikely that Barbara could have lived more than a few more years. For sixteen years, Blalock's shunt had supplemented Barbara's collateral circulation to bring blood past her blocked artery and into the lungs. She died on January 18, 1961. She was twenty-seven years old.

The hope that ran through her family, the miracle they witnessed, and its impact on their lives is still being processed fifty years later. Her brother, a neurologist at the University of California, keeps a photograph of Barbara on his wall and boxes of her memories in his closet. Inside is a 1956 career day speech Barbara gave at her alma mater, Beverly Hills High School, in which she advocated hard work, following a process, and developing one's own ideas. Zeroing in on girls in the audience, Barbara warned against taking secretarial jobs with promises of advancement in favor of work that counts. To follow one's dream, women needed what Barbara called "stick-to-it-iveness." In short: never, ever give up.

*

At the beginning of her career, Helen comforted parents of babies who died because she had no treatment to offer them. Now she comforted parents of pioneering blue babies whose shunts, fifteen or twenty years later, were giving out. Marvin Mason, the third blue baby, was also operated on a second time in Los Angeles and died eighteen months after Barbara.

Helen's blue babies sometimes returned to her when the end was near just as dying rheumatic patients had asked her to visit one last time. When one of her earliest blue babies, a woman in her twenties, lay dying, Helen lobbied a US Air Force base commander to give her brother temporary leave or early discharge so he could be at her side. When an avid skier emerged from a second operation in critical condition, Helen, then seventy-four, visited her daily in the hospital until her death.

She also comforted anew parents whose sorrow was rekindled by her fame. "Dear doctor, I would like to know if it is possible for you to tell me if . . . my little girl had in some way helped other blue babies to live," wrote one Baltimore mother who saw Helen on television in 1966, more than three decades after her child had died. "Let me assure you," Helen responded three days later, "each little blue baby whom I have cared for and followed through the years, either for a long period or short period, has increased my knowledge and therefore each one of them has helped me to help other children." It was by accumulated experience and efforts to help children, to study them, and restudy their precise situation, she told the mother, "that we learned to help so many."[6]

22 CAN SUFFERING BE PREVENTED?

Helen had reached the time in her career when high school girls sought her advice on how to become a doctor and wrote reports on her for school projects. Graciously she answered their notes, offering tips. Requests also poured in from eminent doctors for autographed copies of her new book. The previous year she had published a long-awaited second edition of *Malformations of the Heart*. It had hung around her neck like a ton of bricks.[1] How well her fellows knew! Monday nights in the late 1950s, Helen had asked her staff to stay an hour and listen to her read aloud draft chapters. This was a way she processed information. She had not taken their suggestions, though, and despite his praise, a reviewer for the Royal Society of Medicine was disappointed not to find a fuller discussion of newer diagnostic techniques. Aggravated, Helen prepared a follow-up article.

At sixty-three years old, she was a year and a half from mandatory retirement, vigorously healthy, working long hours. Emeritus, not retired, is how she planned to style herself. She forced herself to think about her successor and long-term studies on her patients. She had already received the most coveted of the twenty honorary degrees she would obtain, from Harvard, likely championed by "Uncle Charlie," Charles C. Burlingham, her father's friend, who repeatedly advanced women for the honor.[2] What a procession, that spring day in 1959! In silk brocade dress, matching jacket, and short white gloves ruffled at the wrist, she walked in a sea of men in black morning coats and top hats wearing a grin as if she had won the break point in a tennis match. She was finally a full professor too, the second woman in forty-two years after Florence Sabin; the promotion squeaked in before the Harvard degree. Helen lectured, mentored fellows and students in her clinic, and saw patients. An expert in suffering, she examined how Vincent Van Gogh depicted it as she wandered the Baltimore Museum of Art one afternoon during Christmas vacation. Life was good, if mundane.

So it was in the third week of January 1962 that she looked forward to dinner in her home with a former fellow who was visiting from Germany for medical meetings in Baltimore. Alois Beuren, "Ali" to colleagues, was older than other fellows. A practicing pediatrician when he read Helen's book on heart malformations, he had been so changed by it that he closed his office and with his wife, Irmie, arrived in Baltimore in 1956 to learn from Helen. In the five years since, Beuren had opened the first pediatric cardiology center at the University of Göttingen and had established himself as Germany's leading authority on children's heart defects. A gregarious man, Beuren was making the rounds of Helen's deputies. On the evening of January 23, he had dined with Catherine Neill and her partner, Lulu Hartarian, on southern-style chicken. He seemed quite cheerful and an account of the evening described festive conversation, with no hint of what he was about to relate to Helen a few days later.[3]

She sat with him in the living room, in stuffed chairs around the fireplace, and offered him a glass of her trademark sherry, prepared to talk about their shared love for Brahms and medical puzzles. Over dinner, Beuren mentioned, almost as an aside, a horrific deformity that several colleagues had seen in their clinics: babies without arms or even legs. "We have seen a very strange malformation in Germany," she would recall him saying. "And what's more, they think it is due to a sleeping tablet." The popular sleeping tablet was no longer sold in Germany, Beuren said. The drug had been voluntarily withdrawn from the market under pressure from public health officials. But the German manufacturer, Chemie Grünenthal, denied that the pill, Contergan (its trade name), had any connection to the misshapen newborns. Its scientists could not reproduce the problem in laboratory animals, and the researcher who raised the alarm was under attack.

Only five weeks earlier, in November 1961, women in several German cities had delivered babies missing arms or legs. At the University of Hamburg, a few hours north of Göttingen, a pediatrician named Widukind Lenz had discovered that half of the mothers there who had given birth to deformed babies had taken a sleeping pill containing thalidomide and sold under the name Contergan. Lenz alerted the local medical community. A week later, on November 15, he met privately with Chemie Grünenthal to warn that the drug was faulty and should be withdrawn. The company's scientists challenged him. How did he explain the fact that some mothers who had taken the drug delivered normal babies or that those who had not taken the drug delivered babies with defects?

Lenz wondered if some patients had forgotten or been ashamed to admit they too had taken the drug. More mothers admitted to taking the drug when he questioned them again, and Lenz forwarded the company additional evidence. He was so worried that on November 20, he alerted doctors at a meeting in Düsseldorf that a drug might be to blame for the deformed babies they were delivering. Afterward, a doctor asked him confidentially to name the drug. The doctor's wife had delivered such a baby and she had taken Contergan. Was that the drug? Yes, Lenz told him. In the next few days, he received letters from more doctors whose own babies had been born with the same horrific defects. He or they shared the letters with government officials, because on November 28, the German Ministry of Health warned pregnant women against taking the sleeping pill, which it identified as the suspected cause of malformed babies. A sleeping pill associated with birth defects was front-page news in Germany. Within one week, under public pressure but conceding nothing, Chemie Grünenthal had withdrawn the drug.

Lenz next telephoned physician colleagues in England, where the drug was also used widely. He alerted the wider medical world in a letter published by the esteemed British medical journal the *Lancet* on January 6, 1962. Lenz came under heavy attack in subsequent issues from doctors who accused him of being alarmist and cautioned against reacting emotionally before the cause of the tragedy could be confirmed. (Replied Lenz, "I can hardly imagine that any person will be able to face the facts without emotion and alarm."[4]) Simultaneously, an Australian physician who had delivered babies without limbs, William G. McBride, wrote to the *Lancet* asking if others had noticed a connection to drugs.

As the evening progressed, Helen struggled to take in Beuren's description of the macabre scene German doctors confronted: babies with stumps for legs, half hands or missing thumbs, even missing ears, and, in place of arms, tiny hands hanging directly from the shoulders, some with fingers. Doctors in ancient Greece had seen babies born with flipper-like appendages for arms or legs and given them the name of *phocomelia*. This condition was rare in the modern world. She went to bed dazed.

When she awoke the next day to freezing temperature and a foot of snow outside her sleeping porch, the mental fog had lifted, and in its place, a heightened tingling that engulfs a person when a path to action becomes clear. Until now, Helen had regarded malformations in the babies she had examined as acts of God. Cells went astray in the early weeks of fetal development, for example. She also knew rubella and other infectious diseases caused changes in the unborn. But here was a drug heralded for its safety in adults—a sleeping pill with no grogginess the next day or possibility of overdose and sold without prescription—that might be maiming unborn children. It had never occurred to Helen—or most others—that a drug deemed so safe for an adult and promoted for the treatment of morning sickness could cause injury to the unborn. If true, it changed everything she knew about how to help children. Could drugs also have caused defects in infant hearts? Throughout her career, Helen had tried to fix these defects, but now she realized she might be able to prevent them. Could suffering be prevented?

By evening, as she sat in a Baltimore symphony hall listening to a concert, Helen decided that the implications of a drug causing birth defects were so critical she needed to verify the cause of the tragedy herself. Or, as she remembered telling herself, "Gracious! I had better go to Germany to find out."[5] Beuren was flying home when Helen wrote letters to his boss, Gerhard Joppich, chief of pediatrics at Göttingen. She was well known there: the university had awarded her an honorary degree in 1960. Joppich urged her to come at once. She then wrote to Lenz at Hamburg, asking if she could review his findings, to investigate for herself.

On Sunday, February 4, Helen phoned Beuren to discuss her plan to investigate. On Monday, she received Lenz's answer: the German medical community had no organized way to study the cause of this horror, and he welcomed her help. By Tuesday, Helen had a plane ticket. She gave herself two weeks to prepare. The chief of pediatrics, Robert Cooke, who himself had two daughters with developmental defects, insisted she not hurry. A crisis of this magnitude could take a month or six weeks to investigate. Neill, the face of Helen's clinic, was an able head.

To Beuren, she named six or eight cities she expected to visit beginning February 18 and asked him to cable her with the name of a Hamburg hotel or, better yet, book a room for her. She offered to see children with heart defects too if pediatric clinics wanted her. Worried about her German—it was poor, she said—Helen cajoled Joppich into finding her an intern from the Children's Clinic at Göttingen to chauffeur her around the country.[6] Her only remaining concern was how to pay for her investigation. If necessary, Helen decided she would use royalties from her book, but the day she booked airline tickets, she also began seeking donations. The International Society of Cardiology Foundation, headed by her Harvard friend Paul Dudley White, sent $1,000 within forty-eight hours. The Maryland Heart Association took more convincing. Heart defects were not common in thalidomide babies, Helen admitted, but the implications were broad and she wanted to learn as much as she could. They donated $1,000; her airfare was covered. Finally, after multiple conversations with doctors at the National Institutes of Health, wherein Helen bandied about White's sponsorship, NIH waived its usual process and agreed to pick up what she estimated would be $4,500 in expenses on the ground.

Aside from her financial backers, her supervisor, and her staff, Helen told no one where she was going or why. Given how the drug company treated Lenz, she did not want to draw attention to her trip before she could investigate the facts. "Don't tell anyone—it is a secret," Helen said when she called her niece, Polly Horn (née Henderson), to ask for a ride from Grand Central Station to the airport. Married and busy with small children, the surprised Horn was hardly aware that her aunt's fame might make her travels newsworthy. In the car, Helen explained why she didn't want a reporter to find out about her travels and show up. "She was afraid to say anything about it because the drug companies were so powerful," Horn remembered. The professors in Germany were in serious trouble, she said Helen told her, and they wanted her to investigate and, if true, to go public with the information "because they couldn't touch her."

Drug companies *were* powerful. They had invented lifesaving antibiotics and psychiatric drugs and doctor-prescribed "wonder drugs" to relieve pain and improve lives. Along with surgery, drugs were one of the major treatment advances of the twentieth century. The public appetite in postwar years for over-the-counter drugs was so great that questionable substances for simple problems like upset stomachs, headaches, and sleeplessness flooded the market. Tranquilizers were hot. Companies sold old drugs at hundreds of times the cost under new names and more alluring advertising slogans. Their stock prices skyrocketed. Helen's colleague at Hopkins would later attribute the phenomenon to "Medicine Avenue," the drug industry's discovery of advertising.[7] This was big business, with a powerful lobby.

And there was nothing the government would do about it. After more than a year of hearings beginning in 1959, US Senator Estes Kefauver, a Democrat from Tennessee, had given up trying to find the votes to impose a federal requirement that drugs work as advertised. Drug companies had resisted regulation, saying the industry could police itself. Kefauver's hearings revealed they had even convinced the medical community.

For centuries, doctors had regulated themselves, and by the 1950s they had a better than half chance of helping people. Doctors charged fees for each service and ran their own insurance programs to help people pay for health care—Blue Cross and Blue Shield associations. If the government began to regulate drug companies, doctors would not be far behind.

The onslaught of drug advertising and price gouging in the 1950s followed a decision by the largest physicians' lobbying group, the American Medical Association (AMA), to close its drug testing office, stop giving its imprimatur on worthy drugs, and, for the first time, allow companies to advertise their drugs in its journal, *JAMA*.[8] The AMA needed revenue: it was gearing up for a fight against another proposed health care plan for the elderly, this time from President John F. Kennedy. In response, Helen had stopped paying dues and canceled her membership.

Helen also knew some doctors promoted drugs without disclosing that drug companies funded them or reimbursed them by purchasing their journal articles and that many of those drugs were useless. A physician at her own medical school, Louis C. Lasagna, had pioneered the so-called controlled randomized design to test drugs, in which he compared a new drug against a placebo and unearthed some worthless drugs. With a small group of pharmacologists, he had started a nonprofit newsletter to publish their drug trials. His work was encouraging other academics to study drugs in a more scientific manner.

Helen did not know, however, that Lasagna had administered thalidomide to forty-one hospital patients at Hopkins and concluded it was a promising new hypnotic, safer than popular barbiturate-type sleeping pills blamed for an increase in suicides. The American distributor, William S. Merrell of Cincinnati, repeatedly stymied in its effort to sell thalidomide, had recently enlisted Lasagna to lobby federal drug regulators to approve the drug.[9] In the United States, the drug was called Kevadon. Merrill was making a final push to get it to market.

*

Helen knew that a small office within the US Food and Drug Administration led by a woman named Frances Oldham Kelsey was hesitant about approving thalidomide in the United States. Her source was one of her former fellows, John O. Nestor, who was now Kelsey's deputy.

The application submitted in the fall of 1960 sat on Kelsey's desk well into the spring of 1961 while she asked for more studies on its toxicity, including after stumbling on a letter in the *British Medical Journal* asking whether the drug was to blame for four cases of neuritis, or numbness and tingling in the limbs, which she realized that Merrell had deliberately withheld.[10] When confronted, the company maintained the reaction was temporary. Kelsey was not appeased. If the drug caused nerve damage in adults, she wondered whether the fetus too would suffer nerve damage.

Reserved and self-effacing, Kelsey was tougher than most people realized. She enjoyed telling how she had been accepted into the doctoral program at the University of Chicago because Eugene Geiling, the pioneer of drug testing, assumed from her name that she was male.[11] One idea she worked on in his lab was whether drugs might cross the placenta, the organ that develops in the uterus to allow oxygen and nutrients to pass from mother to fetus and filter out waste. She performed drug tests on animals and helped show how a mass-marketed drug had killed people. She also knew from a part-time job at the AMA while earning her medical degree that doctors sometimes promoted drugs in which they had a financial interest.[12] The FDA rules she had to work with were weak. Companies had only to present general testimony that a drug was safe, and if the government did nothing, the drug would be approved in sixty days. After the worthless drugs described at the Kefauver hearings and criticism from Minnesota senator Hubert H. Humphrey, Kelsey and other scientists had vowed to examine new drug applications more rigorously.

Kelsey had already found Merrell's application incomplete five times when the frustrated company sought a meeting at the FDA on September 7, 1961, to which they brought Lasagna to pressure her and her bosses. They wanted thalidomide on the market by mid-November. Lasagna said the drug was worth trying despite "hangover" effects. He was skeptical that it was safe at higher doses, though. And when Kelsey suggested it would be wise to test its impact on the fetus, he agreed.[13]

The company scrambled. Without telling Kelsey, they sent the drug to more doctors to try on patients and convince her of its safety. Among the doctors were several hundred obstetricians and gynecologists, doctors with direct access to pregnant women.

In November, Kelsey held it back again, this time over its label, forcing the company to wait another sixty days, until January 1962. Kelsey wanted the label to state that some neuritis had resulted in patients and because neuritis could affect the fetus, the drug was not safe for pregnant women. Then Merrell, the US distributor, learned that the drug's German manufacturer had withdrawn the drug in Germany. When Kelsey learned that malformed babies were involved and again confronted Merrell, the firm told her thalidomide could not be the culprit because no babies with defects were born in the drug's early years.

A week later, on December 4, Merrell told US doctors testing its drug for neuritis that although there was no direct causal relationship between the drug and malformed babies, to be on the safe side, it should not be taken by pregnant women or women of childbearing age. But it still hoped to win Kelsey's approval. In mid-January, Merrell even enlisted a Philadelphia drug company to reassure worried investigators, telling them that Lenz admitted he had failed to directly link malformed babies to thalidomide. Thalidomide had been a best seller in Europe, and even without women of childbearing age, the market for the sleeping pill in the United States was huge. Its defenders included British doctors who used it on children with epilepsy and as a tranquilizer for uncontrollable children, some with developmental disabilities.

*

When she arrived in Hamburg, Helen found a beleaguered Lenz and his wife appreciative of her support and skills. By then, Lenz and his colleagues were delivering between three and ten babies a day with missing or deformed parts.

Helen pored over Lenz's patient files, which he was scrutinizing for more evidence to buttress his claim that the drug was to blame. Why were some babies missing arms and legs and bladders, but others missing an ear or a thumb? Working side by side with Lenz for days, Helen saw in exacting detail how he dug deep into his patients' lives; he examined prescriptions, read hospital pill bottles, and reinterviewed mothers with a calendar in his hand to confirm that women who had given birth to babies with missing or malformed arms or legs or noses had taken the drug on the precise days their fetuses were developing arms or legs or noses. The more Lenz learned about when mothers took the drug, the easier it became to predict which of his patients would deliver malformed babies and the type of defect based on the date in their pregnancy they had taken the drug.[14]

After a week in Hamburg, Helen was ready to expand her fact-finding. The young pediatrician who volunteered to accompany her, Hans J. Severidt, met her in Düsseldorf with a car and a list of contacts. They spent between one and three days at a time in Frankfurt, Heidelberg, Freiburg, Munich, Göttingen, Bonn, and Stolberg, interviewing pediatricians and researchers, examining babies in their clinics, taking notes on their symptoms, and cataloging dates and doses of Contergan their mothers had taken. From doctors she interviewed, Helen learned of parents who could not stop weeping after seeing their babies. Revulsion was common. Many screamed. Some mothers hoped their babies would die, and those without bowels did die. But two-thirds of the babies lived. Their brains were normal, their smiles and cooing as sweet as any baby, and it was this incongruity that struck Helen as unconscionably cruel. She kissed and tapped them affectionately on the cheek the same as she kissed her blue babies. She felt for their heartbeats with her hands. She remembered what it was like to be held back by a physical limitation, as she had been. It would be far more difficult for children without arms and legs to fit in.

"They are ghastly children—many with no ears, commonly and the worst, no arms—rudimentary hands from the shoulders—stumps of legs—other anomalies of the G.I. tract and heart," she wrote Neill.[15] What most upset her were the infants who could not turn over in their cribs or swallow on their own. Many thalidomide babies died within the year, some from parental neglect or intentionally administered drugs. It was heartbreaking work. "I am seeing one of the terrible medical disasters of all time," she wrote. Photographs could not convey what Helen experienced when examining these children in the presence of their mothers, she said. But the images would go far to explain the impact of untested drugs. She asked Lenz for copies and permission to distribute them.

*

Combining Lenz's findings with her own, Helen made charts for babies
with ear deformities and compared them with the dates their mothers had
taken the drug. It was always between the 35th and 37th day of pregnancy.
Heart problems could be traced to pills taken on the 39th day. Arm deformities showed up in children whose mothers took the drug between the
39th and 41st day of pregnancy, and missing legs were linked to Contergan
consumed between days 41 and 44. These were critical days in the development of the fetus. It appeared that any malformation and the type of
malformation that occurred depended on when the mother had taken the
drug. The cases Helen found confirmed Lenz's theory.

In Heidelberg, Helen stopped at the US Army Headquarters to verify that
Contergan was never dispensed at the base and there were no abnormal
pregnancies. In Bonn, she spoke with a doctor who documented ninety
cases of phocomelia in babies whose mothers had taken the drug during
early pregnancy. Another doctor told Helen that in his fifty years of practice, he had delivered a baby with flippers as often as he delivered one with
two heads: never. Her notebooks filled, Helen was 90 percent convinced
that thalidomide was the cause. At night in her hotel room, she sketched
the chemical composition of thalidomide, trying to understand how it
worked. Did the drug cross into the placenta? If so, how?

Back in Göttingen, she and Beuren puzzled over why the drug did not
seem to affect the offspring of pregnant animals in laboratory tests. Dozens
of German researchers had begun testing the drug. Beuren gave his pregnant
dog large doses of Contergan and promised Helen he would report back.

From Germany, Helen flew to England, the second largest market for the
sleeping pill. She met her sister Mary in London and the pair relaxed for a
day, dining at the Radcliffe Club. She thought not only about the cause of
the defects she was seeing. She also tried to imagine the lives these children
faced and how pediatricians would care for them. Those she interviewed
struggled with the best approach. In London, she had examined an infant
girl with fingers hanging where her arms should be. The girl's doctor suggested amputating her fingers. Horrified, Helen made a note to find an
expert on orthopedics to help this infant.

In England, thalidomide was sold under the name Distaval and marketed, as in Germany, as an antidote for morning sickness. The British

distributor, Distillers Limited, a subsidiary of the liquor manufacturer that had produced drugs in wartime, had withdrawn it from the market in December following action by the German manufacturer. After this experience, it would return to its base business, scotch. Charles Brown, a deputy of Distillers's chief scientist, George Somers, readily shared his lab investigations. He had administered the drug to pregnant rabbits and replicated the abnormal births. He also shared that the drug was off the market in Australia too.

Helen was now "100 percent convinced" that the anomaly in babies was caused by thalidomide. In Brown's lab, the drug had been shown to pass through the placenta, from the mother to embryo or fetus. Helen was consumed by the possibility of similar effects from other untested and easily available drugs. Instead of a salve for women of childbearing age, potential harm lurked in these drugs. She had a mission, "quite a mission," she told Charlotte, to educate other doctors and change the law.[16]

Apart from an essay in *Time* magazine, "Sleeping Pill Nightmare," on February 23, 1962, the American press ignored the thalidomide disaster in Europe. If a news organization had tried to write about it, it would have run into Merrell lawyers or executives. They argued that since there was no "scientific" or laboratory evidence tying the drug to malformed infants, claims or stories that sought to draw such a relationship represented an emotional and unfair attack against drug companies. They accused *Time* of bias.

Helen knew what she was up against. Lenz, so careful with his facts, had been challenged, bullied even, by the chemical company and in the *Lancet*. Some German researchers theorized that even if thalidomide was to blame, it might be only one factor. She would need to be sure of her facts to avoid being entangled in legal disputes or damage her credibility. But she was confident she had overwhelming circumstantial evidence from multiple locations that linked the drug to birth defects, and she planned to reveal it. Of all the players in this drama, she alone had international credibility on infant malformations. If she raised the possibility that other drugs could have similar effects, people would listen. On the plane from England to Baltimore, she began her to-do list.

*

To Helen, thalidomide demonstrated the need for exhaustive testing of new drugs on animals before any human trials. In a report to Paul Dudley White, one of her financial backers, Helen described the "horror" she had seen and her plan to strengthen US laws regulating the sale of drugs to the public. "It certainly is only by the 'grace of God' that our country has escaped," she wrote.[17] To her dismay, Helen found that thalidomide was sold by distributors around the world under dozens of names. Until she knew them all, she could hardly warn unsuspecting women against take the sleeping pill. Her first act was to write to former fellows and colleagues with a plea: Please tell me whether this drug is sold in your country, under what name, and in what dose. Have you any reports of phocomelia babies?

In 1962, a letter took three or four days to cross the country and up to seven days to traverse continents. Zip codes would not be introduced for another year and overnight mail another decade. Every professor had a secretary to type for him or her from dictation tapes or handwritten notes, but tapes could be lost and notes misread. Letters were typed in triplicate, the copies on flimsy paper with smudged ink and the initials of the typist on the bottom. The Xerox office copier, invented in 1959, was being installed in big corporations.

The tools Helen used to gather and communicate what she had discovered in Germany would seem primitive in a decade's time. She had slides and photos of babies, pages of her own handwritten notes, some typed notes from Lenz, and her own drawings. She had typed tallies and columns of numbers of patients, names of drugs, dates during pregnancy a drug was consumed. Her mathematical skill was as fierce as her memory. There were no computers. She calculated by hand, three times, as was her custom.

In a dark green spiral notebook five by seven inches wide, she made columns of lists, for countries, distributors, and thalidomide's street names. While she waited for data, she sat at her desk and outlined a talk and a paper summarizing her findings and her experiences in Germany and England. She detailed the condition and why she suspected thalidomide as the cause. This would become the basis for her campaign.

*

The day after she returned from Europe, in the first week of April 1962, Helen presented her slides and research to staff, Hopkins colleagues, and doctors she invited from the nearby University of Maryland School of Medicine. She described missing limbs, gallbladders, and partial digestive tracts.

The fifteen photos she showed, babies with flipper-like arms, tails in place of legs, and missing ears, stunned the savvy audience.

In the next few days, Helen convinced colleagues at two major medical groups to give her time at their upcoming spring meetings. The typical route for announcements, publishing a paper, took months. She told them it was an emergency. By Wednesday of that first week, Helen reconnected with her former fellow and source, Nestor, at the FDA. He had heard about her trip during a visit to Baltimore, and he wanted to know what she had learned. By Friday, April 6, 1962, Nestor and Kelsey were in Helen's office. That they called on Helen indicates how eager they were for more information to bar thalidomide from the marketplace and how busy Helen was making plans to publicize what she considered a health emergency. The FDA had no power to ban drugs even in the testing stage. With no information of its own, it could not even warn people. That, Helen already knew, would be up to her.

23 IF THEY HAD SEEN WHAT SHE HAD SEEN

Helen emptied her notebooks to show the connection between thalidomide and phocomelia babies in her meeting with Kelsey. Babies with congenital deformities that had been mistakenly attributed to genetic disorders in 1959 had by now been tied to their mothers' use of thalidomide.[1] Thereafter, the number of babies born with defects increased geometrically until late 1961, when Lenz in Germany, Somers in England, and McBride in Australia simultaneously linked the condition to a drug and sales were halted. Groups with easy access to this drug, the wives of doctors and the wives of Contergan salesmen, had delivered babies with missing or malformed parts in disproportionate numbers. Women living on US Army bases in Germany, where Contergan was not available, delivered normal babies.

It was worse than Kelsey imagined. The information Helen provided, her charts from twenty clinics in Germany and England, her calculations showing when women took the drug and the types of babies they delivered, was enough to bar the drug from the US market. How many other new drugs could be dangerous to women and children? Helen told Kelsey that since women often didn't know they were pregnant when they took medicine for routine illnesses, all new drugs should be tested on animals for their impact on the unborn. But Congress, not Kelsey, would have to be persuaded to change the laws to give regulators more tools to evaluate proposed new drugs. Helen offered Kelsey her data and told Kelsey she expected to be called as a witness to support tougher drug testing. Meanwhile, she would alert doctors of the danger to women of childbearing age.

There was one more thing, Helen told Kelsey: since the drug had been distributed in the United States for investigational use, at least some malformed babies would be born to American mothers.

Initially, Merrell had given the drug to 37 doctors who tested it in 1,589 patients. When Kelsey discovered some patients experienced neuropathy and asked for further testing, she had asked Merrell to update her. The company was vague; it said only that it had received results from 56 physicians. The patients numbered 2,961. Helen suspected there was a lot more thalidomide out there. She told Kelsey to be sure those drugs were destroyed before they reached more pregnant women.

Kelsey and Nestor left Helen's office that Friday in a hurry.

*

Early the next week, Kelsey wrote to the Cincinnati distributor seeking an updated list of drug investigators and what Merrell had done to inform these doctors about the drug's potential harmful effects on the unborn once they learned of it.[2] Meanwhile, Helen took a train to Philadelphia to corner doctors on the last day of the annual meeting of the American College of Physicians. People were leaving when a call came over the loudspeaker for a special session, a talk by Helen Taussig on the phocomelia outbreak in Germany. Helen's medical school housemate, Helen Pittman, turned around. "I thought, I ought to stay and listen," she recalled.[3] What she heard on Wednesday, April 11, 1962, was the first public warning in the United States about the danger of thalidomide tablets sitting in doctors' offices and the potential harm from *any* drug on the market to the unborn fetus.[4] After describing babies born to mothers who had taken the drug as "the most ghastly thing you've ever seen," Helen asked doctors to alert their patients.

Helen paid tribute to Kelsey, saying thalidomide was not offered in the United States because of Kelsey's suspicions about the drug's potential impact on the fetus from an unrelated problem, neuritis, a painful tingling of the nerves. How easily it could have been approved, she said, given the lack of a requirement for companies to show that a drug worked. She called for testing all proposed new drugs for their impact on the unborn.

Helen's talk sent chills down the spines of pharmaceutical representatives in the audience. This was a tougher standard than in the Kefauver proposals they had defeated in the past and that Kefauver had just reintroduced. In an editorial the next day, the *New York Times* called for better drug safety laws in response to Helen's shocking tale. "People are not guinea pigs, and the law should not permit them to be used as such," the paper said.[5] Four days later, the chief of neuropharmacology at Abbott Labs, Guy M. Everett, asked Helen which animals she thought could be tested, since extensive tests of pregnant animals had already been conducted, with no abnormal results. She told him to try again.[6] She herself was about to try to replicate British researchers' tests on rabbits. In pursuit of "absolute" or direct proof to tie the drug to malformations, Helen had enlisted an expert in embryology at the Carnegie Institute in Baltimore and given him some of the drug to test on white rabbits.[7] Either Helen financed his lab tests or he performed them gratis, because she failed to cajole NIH officials into letting her repurpose the $3,000 remaining from its $4,500 thalidomide travel grant. (They did thank her for being frugal.)

But on the strength of her circumstantial data alone, for the next eight months, from April through November 1962, Helen single-handedly waged a campaign for tougher drug testing laws to protect women and children. In her office, in hallways, on rounds with interns and residents, on the phone, in meetings with doctors, journalists, and government regulators, in speeches, in hundreds of detailed letters to scientists, parents, and corporate executives, Helen explained thalidomide and its repercussions. In Atlantic City that May and at medical conferences in other cities in June, July, and August where Charlotte met Helen, and whenever the two spoke by phone, the subject was always thalidomide. "At that time she couldn't think of anything else," Charlotte recalled. Unwilling to hear any more, Charlotte skipped her annual trip to Cotuit.

*

Warning doctors at meetings as Lenz had done was only the beginning. Helen's effort to alert the medical community at large through *JAMA*, the country's most read medical journal, led to a head-on clash with John Talbott, the editor. *JAMA* insisted on exclusivity. He refused to publish her warning letter because *Time* magazine had already written about the German disaster. Helen was offended. The *Lancet* had immediately published letters raising ominous questions about thalidomide. "I honestly feel that we have more of this drug in this country than we realize and the danger of severe malformations is great," she told him.[8]

Helen's pushback worked. *JAMA* had to be sensitive to rejecting criticism of a drug after the AMA disbanded its drug testing office. Knowing Helen would find other outlets or perhaps because he was genuinely interested, the editor invited her to pen a major article. Instead of a warning letter, he wanted more science, more facts, something exclusive. He wanted all her evidence. *JAMA* was competing with medical journals beginning to publish more research rather than one doctor's report. This was an important opportunity. Immediately Helen began drafting.

The price was more than a month's delay in publication. While she worked on an exclusive for *JAMA*, Helen urgently alerted doctors to the danger of samples that might be in their offices in other journals that were not so stuffy. In early May, *Science* published her letter warning about thalidomide and the potential harm to the fetus from all new drugs.[9] The *American Journal of Diseases in Children* published a similar letter in June. The *New England Journal of Medicine* referred to her findings when it suggested that medicine could be practiced better and more safely with fewer drugs and "a greater reliance on the healing properties of *vis medicatrix naturae* [the power of nature to heal the body]."[10]

After her April talk in Philadelphia, Helen became the clearinghouse for information and advice on thalidomide and other drugs. When the first American thalidomide babies were born—twins to a German mother in New York—she began advocating a national registry independent of the drug industry to report such births. When a California public health official sought her advice on the polio vaccine—the Sabin live vaccine was being tested—Helen began warning in talks and letters against giving live vaccines to pregnant women for fear they too could cross the placenta. She cited the effect of German measles and thalidomide on the fetus. When a New York doctor asked if he should still immunize young women traveling on their honeymoons to tropical lands, she said no. "It is a bad thing in light of our present knowledge," she told him.[11] She would be proven right.

The second physicians' group Helen addressed within weeks of her return was pediatricians. Many children's doctors who heard Helen speak in Atlantic City in early May 1962 came away with heavy hearts for child victims.[12] Some wanted to do something. As Charles A. Janeway, chief of Boston Children's Hospital, would tell Helen, the number of powerful drugs that had become available in thirty years had changed medicine beyond recognition.[13] But whether and how to test drugs on children was a growing debate with little consensus among pediatricians.[14] Some children's doctors, like drug company executives, questioned the value of Helen's proposed "early warning system" for congenital malformations. How effective could a test of new drugs in pregnant animals be, one doctor asked her, if thalidomide did not produce malformations in animals? Another suggested consulting researchers on toxicity.

Her response echoed what she had told colleagues who questioned her study of babies destined to die from heart defects. If the goal was worthwhile, it was worth pursuing. Eventually someone would figure it out. She likened the quest to President Kennedy's promise to land a man on the moon. Alan Shepard had just orbited Earth in the spacecraft *Freedom 7*.

Unlike the AMA, the American Pediatric Society did not have a stated position in support of drug industry self-policing. Helen proposed they form a committee to prevent another such tragedy. On her own, to build support, in June 1962 Helen penned a long letter to pediatricians asking them to lobby their congressmen; by return mail, using a form she provided, dozens of prominent doctors urged her on.

Merrell, the marketing arm for Richardson-Merrell, formerly Vick Chemical Co., the cough-drop company, did not attack Helen as Grünenthal had attacked Lenz. She had not yet revealed her evidence. By this juncture, she had data on more babies whose deformities tracked dates their mothers consumed Contergan than any physician and, most assuredly, the drug's manufacturer and its lawyers. At Helen's request, Merrell even supplied her with thalidomide for the animal tests she had begun. She kept a canister of it in her office. But US pharmaceutical companies, which conducted their own drug tests, recognized the enemy and began a counteroffensive. They argued that the current regulatory system worked because the drug had never made it to the US market. Quickly, executives of top firms formed a new industry council and told Helen and members of Congress that if better testing were needed, their new group would be the best overseer. One of the industry's long-time defenders, Morris Fishbein, a former editor of *JAMA* who for a quarter century helped set the policies of the AMA, argued that government regulation would impede the ability of reputable pharmaceutical companies to develop and win approval for lifesaving drugs. He labeled witnesses who had testified before the Kefauver committee discontents and "scientific leftists"—code in the Cold War era for "Communist."

He was gentler on Helen. Whatever her scientific credentials, he argued, she was not an expert in drug research or development.[15]

Aside from the *Times*, the lay press ignored Helen's warning. She had no thalidomide patients to demonstrate the tragedy, the way she used blue babies whose lives had been saved as exhibits of the benefits of experimental dog surgery. So when the chair of the House Judiciary Committee, Emanuel Celler, Democrat of New York, asked Helen to testify on thalidomide before his antitrust subcommittee on May 24, 1962, she agreed on one condition: that she could show Lenz's photographs of thalidomide babies. She did not show the worst ones. But among the photos of smiling and bright-eyed infants, one baby had no arms and another had flippers for arms. Helen made it clear that neither she nor the congressmen needed a pharmacology degree to see what she had seen and draw similar conclusions.[16] If these babies were their own children or grandchildren, Helen asked, "Would you not exert your influence to prevent the occurrence of another similar tragedy?" At minimum, she said, Congress should force manufacturers to list side effects on a drug's label in the same-size typeface as indications. It was particularly important for commonly used drugs for colds, allergies, or sedatives. Above all, drugs should be tested for their safety in infants and children. "Safe should mean safe for all ages and for all groups of people," she said.[17]

Her testimony led the editor of *Scientific American* to commission an essay by Helen about her European trip. It was small solace when news outlets ignored her photos, and, once again, drug industry supporters in Congress secretly scuttled tough provisions in a Kefauver bill making its way through committee. Helen admitted to being "desperately worried" her campaign would fail. She could not even convince doctors to support reform. "If we had a thousand cases here, they would all be up in arms. . . . It may be a long hard fight," she told a colleague. "I don't know where to go from here."[18] But she had already set in motion a story that would galvanize public opinion and push Congress to act against dangerous drugs.

*

In all her pleas for tougher drug testing, Helen promoted Kelsey as a smart scientist who had staved off the drug despite weak laws. She pointed out how easily thalidomide might have been sold in the United States in the hands of a less careful regulator. Only luck and the right person in the job had prevented the tragedy from crossing the ocean. Kelsey was the "the Grace of God" she talked about. Nobody but the pharmaceutical companies paid attention. Part of it was indifference to disaster in other countries. Congressmen upset about price gouging in the drug industry did not care about a government regulator stalling a drug that caused tingling in the arms. That changed with Helen's testimony. Her display of disturbing photos revealed exactly what Kelsey had prevented in the United States. The evidence Helen had gathered to blame thalidomide for birth defects in Europe was also about to be unveiled in *JAMA*. Now, suddenly, Kefauver's staff saw in Kelsey the story of an American heroine. In a last-ditch effort to revive a bill to regulate the drug industry, they tipped off a *Washington Post* reporter to the story of a government regulator whose suspicions thwarted a tragedy of Shakespearean proportions in their own country.[19]

While the *Post* pursued the story, Helen, with help from Park, tussled with a rewrite of her *JAMA* article. In a "special communication" to readers on June 30, 1962, *JAMA* published Helen's nine-page article replete with charts and photos and a drawing of the chemical.[20] The case was circumstantial, and Helen approached it with the skill of a prosecutor, acknowledging the weaknesses, letting the narrative unfold as a kind of mystery, carefully releasing one powerful detail after the other. Some German doctors were doubtful, true, and little was known about how the drug worked, although Grünenthal had confirmed to Helen that it passed through the placenta. Few animal studies had been done, and the results were mixed. A British scientist's study of the drug in rabbits resulted in abnormal births, but the Germany manufacturer's test on rats did not. So far, German researchers had ruled out X-ray exposure, hormones, detergents, food preservatives, and contraceptives as potential causes.

But a few German doctor-investigators were so convinced of the link that they went to extraordinary lengths to verify their patients' recollections about taking the drug. Lenz at first found 20 percent of patients with malformed babies had taken it, but after additional detective work, the number grew to 50 percent. In one case in which the mother's physician swore he had prescribed a different sedative, Lenz located a record stating that the sedative was not available and Contergan had been given instead. In the case of a mother who denied taking the drug, Lenz found a hospital had mixed thalidomide into a sleeping pill given to her for minor surgery.

When he could not obtain hospital scrips or other documents as proof, he accepted dates from mothers who linked taking the drug to specific events, such as the night a neighbor's house burned down or three days after a father-in-law had been murdered.

City by city, Helen illustrated how the findings of doctors she visited confirmed or complemented each other. One doctor in Düsseldorf found no cases in three hundred women who had *not* taken the drug. Half of the women who *had* taken the drug delivered abnormal babies. In another group of forty women who had taken the drug, 75 percent had abnormal babies. Reports from around the world were still coming in, but so far Helen documented abnormal births to mothers who had taken the drug in Sweden, Switzerland, Canada, Brazil, Peru, and Canada.

She estimated a minimum of four thousand babies would be born with tragic defects because their mothers had taken the drug during critical weeks of pregnancy, and one-third of these infants would die. On a desk in Westphalia, she had seen a stack of records waiting to be entered to a registry for children who would need orthopedic aids; eight hundred had already registered. She also explained why some pregnant women who had taken the drug delivered normal babies. They took it in the first few weeks of pregnancy or after the first two months, making clear that the drug acted as the embryo developed. The only light in this dark tale was that it opened research into the causes of malformations and fetal development, Helen wrote. This was the dawn of a new body of knowledge, a paradigm shift in the understanding of human development and the physician's role in promoting health.

Eleven days later, on July 11, Senator Kefauver reintroduced a bill to oversee the drug industry. Fifteen days later, a front-page story in the *Washington Post* by Morton Mintz told how Kelsey had stalled the drug company multiple times to obtain more information about thalidomide's safety.[21] Her efforts to obtain information on the drug's effect on the nervous system and delay approval were unusual and creative. Her immediate boss backed her demand for state-of-the-art drug tests despite higher-ups who invited drug companies to visit the FDA. The story showed that the rules in place to approve drugs were weak: only a two-month window for government to approve new drugs, testing of drugs in doctors' offices without adequate or any warning to patients, and little or no consideration of a new drug's impact on women of childbearing age. The government lacked the power to ban unsafe drugs. It relied on the manufacturer's honesty.

The connection between the two women's work was apparent. Kelsey had avoided a crisis because of her questions over the drug's impact on the fetus and exposed the flimsy wall protecting Americans from greedy companies. Helen had gathered evidence that the government could not, and private companies would not, in order to reveal the drug's potential danger. The *Post* quoted Helen and cited details from her *JAMA* report to reveal how Lenz in Hamburg and Helen in other cities linked the drug to malformed babies.

The *Post* story inflamed public opinion. Many doctors participating in the test of thalidomide had no idea they had been engaged in human trials exposing patients to harm. Their patients did not know either. Adding to the furor was the discovery that samples of thalidomide were missing. In July 1962, three months after Helen had warned Kelsey about the danger from test tablets and begun alerting doctors, the FDA launched a campaign to find drug samples. Helen was right: there was a lot more thalidomide out there, more than any other unapproved drug in history. Instead of 56 physicians and 2,961 patients, the distribution numbers were closer to 1,200 physicians and 20,000 patients. In hearings chaired by Senator Hubert Humphrey in summer 1962, the FDA estimated Merrell had distributed 2.5 million thalidomide tablets.

*

Helen might have stopped at this juncture, leaving the political system to respond to the public outrage she helped generate. Instead, she intensified and expanded her lobbying, weaving outside her academic lane in ways that left some colleagues uncomfortable regardless of their views on drug test reform. She believed it was her duty as a physician to take her campaign directly to women of childbearing age.

A serious-looking Helen, in glasses and white coat, admonished readers against feel-good but potentially dangerous drugs in a *New York Post* Sunday magazine article, "The Pill that Cripples." She even extracted a promise from the reporter to help her explore why healthy women succumbed to advertisements for pills.[22] The question haunted Helen. If a woman needed sedatives to get through pregnancy, how good a mother would she be?

She targeted women's magazines—*Ladies' Home Journal*, *Glamour*, and *Redbook*—to push her message against taking drugs during pregnancy. Next to Helen, Kelsey sounded calm when she explained to *Ladies' Home Journal* readers that new drugs would always have unexpected effects and listed four that had been withdrawn by their makers because of patient death or injury. They included Mer/29 or Triparanol, a cholesterol-lowering drug distributed by Merrell that left patients with permanent cataracts. "Why are women so casual about drugs?" Helen asked in *Glamour*, which explored the question in October 1962. She blamed advertising and admitted she'd been called old-fashioned for not knowing about the "newest model wonder pill" rather than waiting to see if it was safe. She was a naturalist, after all; she ate dog meal, for example, to speed the healing of a broken leg (when niece Polly saw her eating it surreptitiously in the kitchen, Helen asked her not to mention it to visiting doctors.) She had even withstood the most popular product marketed to women in the 1950s, hair spray, whose chemicals, it turned out, affected the environment and women who used them. By the end of a busy week, Helen's usually coiffed hair poked out of a net.

Helen also wooed general interest media, writing to Norman Cousins, the editor of the *Saturday Review*, which had exposed deceptive advertising and crusaded for tougher drug regulation beginning in 1959. She would help the magazine's science reporter, John Lear, who was preparing another series of articles on drugs. The *New Yorker*, which had just published excerpts from Rachael Carson's *Silent Spring*, revealing the danger of chemicals and igniting the environmental movement, rejected another essay that revealed similar dangers in drugs. ("Dear Dr. T: Not at this time," came the short, kind reply.)

Throughout July, as she pushed for media coverage and gave interviews, Helen lent copies of Lenz's photos of smiling babies lacking legs or arms or with tiny fingers hanging directly from a shoulder. *Redbook* rejected a story, telling Helen her photos would terrify prospective mothers. "As if it isn't far better to scare them than to have a deformed infant," Helen complained to a Princeton University professor whose publisher also stripped the photos from his book on prenatal care.[23]

Scientific American too refused to publish photos with Helen's essay for the August edition, over her protests, calling them too grotesque. Instead, the editor commissioned drawings. Helen's seven-page essay in *Scientific American* hit newsstands in late July 1962 and established her as a leading advocate for government protection of consumers. It was a layperson's version of the *JAMA* piece, told in dramatic form, with an ominous description of how drug industry practices led people into taking poorly tested drugs. In welcoming style, Helen brought the reader along as she investigated an allegedly safe drug taken by thousands of women who delivered monster-like babies.[24] *Don't blame us*, the drug manufacturer had said, pointing to clean laboratory tests. But Helen's readers, following her search for evidence, could see for themselves the notebook full of dates when pills were taken and the corresponding deformities in infants.

This time she described drug company sales practices, a topic not mentioned in her earlier piece in *JAMA*. She explained how a generation earlier, new drugs only slowly found their way to popularity based on doctors' experience and knowledge. Now, drug company salesmen were visiting doctors' offices and "assaulting" them with fancy brochures and placing glossy color advertisements in medical journals that extolled the latest iteration of a simple drug, say a diuretic, tranquilizer, or antihypertensive compound, with the result that these new drugs found markets within months. Helen also let readers in on how experimental drugs were distributed to patients without warning or tests on animals.

For some readers, her tale evoked experiments on Jewish prisoners in Auschwitz, the Nazi concentration camp. Helen did not publish all she gleaned from Merrell documents her colleague Lasagna shared. In Germany, thalidomide was tried on children who did poorly in school, the senile, the infirm, mentally ill, and 370 patients at a gynecology clinic. Such tests would not seem to respect the principles of inform-and-consent for human experimentation established in the Nuremberg Code in the aftermath of World War II. In the United States too, the drug was tried on chronically and mentally ill patients as well as three-year-olds afraid of their dentist.[25] Hospitals and doctors at the behest of drug companies regularly tested drugs on unknowing patients.

*

Helen's *Scientific American* essay in summer 1962 coincided with a raging debate in Europe over mercy killing. The debate was prompted by the trial of a Belgian mother charged in the murder of her thalidomide baby. Suzanne Vandeput Coipel was arrested after adding a lethal sedative to the infant's bottle. She, her husband, and her doctor would ultimately be acquitted. In London, Lady Edith Summerskill, a physician, suggested that abortion should be permitted in the case of a woman who had taken thalidomide during the critical weeks of pregnancy. British laws, like those in the United States, allowed abortions only when pregnancy endangered the life of the mother. At the end of July, a British National Opinion Poll showed almost 73 percent in favor of abortion where there was a good chance the baby would be deformed.[26] In Cotuit, Helen investigated the case of a US-born phocomelia baby. When she finally stole away in late July to fly-fish with Park at his remote cabin in Cape Breton, her deputy found her there nevertheless. The question of whether a woman should be allowed to abort a baby conceived while she was taking thalidomide had reached America. Doctors and reporters wanted Helen's opinion.

In Phoenix, Sherri Finkbine, the mother of four and the local host of the children's TV show *Romper Room*, privately sought permission from a hospital medical committee for an abortion. She had taken the drug early in her pregnancy while in England with her husband. She was two and a half months pregnant. Her doctor assured her she would have no problem obtaining the abortion; as required at that time, she had two statements from psychiatrists stating that her life was in danger. She said she would kill herself if forced to have the deformed baby.

On the weekend before her scheduled abortion, Finkbine told her story to the local paper anonymously, with the goal of warning women not to take drugs during pregnancy. The subsequent front-page story delayed hospital legal review. There could only be one such woman in Phoenix, and instead of routine approval for her abortion, the hospital balked. Finkbine took her case public when on Friday, July 27, a day after she was to have had an abortion, she and her husband filed suit seeking a legal opinion that she was entitled to the procedure. They wanted assurance she would not be prosecuted. The judge refused. Under existing law, if hospital officials decided her life was in danger, she could obtain an abortion. Wednesday night, July 31, hospital officials refused Finkbine an abortion. With her children often at her side and her simple statements about the difficult choice she faced as she pursued her quest, Finkbine cut a sympathetic figure and changed American attitudes toward abortion.

*

As the foremost US expert on the impact of thalidomide on the fetus, Helen was the one to consult on the prospect of a normal birth to a woman who had taken the drug in pregnancy. With Neill as intermediary, calling Helen in Cotuit to read incoming cables and take notes to respond on her behalf, Helen privately dispensed treatment options in late July and early August 1962 to doctors with patients like Finkbine.

"STOP in my opinion therapeutic abortion is indicated," she cabled a doctor in New York City who had described the dosage and the dates his patient had taken the drug.[27] To another doctor whose patient had taken a smaller dose over a shorter time period, Helen wired, "Normal baby but with saddle nose!" The word *hope* was scribbled on one cable in which Helen appeared to be recommending abortion. Showing her determination to help women in dire circumstances, Helen even asked a colleague whether he would perform an abortion for a woman who had taken thalidomide during the dangerous period and could not obtain the procedure elsewhere. There was no answer.

The patient Helen advocated for now was the mother, and she was unafraid to do so publicly, eschewing a science reporter's offer of anonymity in exchange for her opinion on a subject as controversial as abortion.[28] She told Neill to give the reporter her number in Cotuit. Not every pregnant woman who took the drug delivered a malformed baby, she explained; it had to be taken at a certain time in the first two months of pregnancy. Each case was unique. Throughout the summer, Helen reiterated her belief that abortion was preferable to delivering a malformed baby. "A therapeutic abortion for the woman who has taken thalidomide is quite as thoroughly indicated as it is for a woman who has had German measles in early pregnancy," she wrote Lear of the *Saturday Review*.[29] She called it a delayed pregnancy. The mother could have another child. Lear's influential articles about the drug industry had begun appearing in July and August 1962. After one lengthy phone call in early August, she airmailed him a three-page, single-spaced letter on how drugs should be regulated. She included the photos from Lenz. From Cotuit, Helen filled in Charlotte on her schedule, including flying to New York City on the eve of a July 26 interview on CBS.

*

Drug industry friends in Congress appeared poised to prevail. Support for Kefauver's bill teetered in June over its provisions to outlaw industry collusion and price fixing. Yet public support for drug safety was building. On August 1, with television reports on Finkbine's effort to obtain an abortion, newsstands touting Helen's *Scientific American* description of drug sales practices, and the FDA searching for thalidomide tablets, President Kennedy announced his own compromise bill. To the dismay of the bill's sponsors, it dropped many antitrust features opposed by industry. But it required government approval for controlled tests of new drugs and gave the government authority to immediately remove flawed drugs from market.

Ten days later, on August 10, under the glare of worldwide publicity, Sherri Finkbine and her husband flew to Sweden to obtain an abortion. In *Life Magazine* that morning, she was photographed in her bedroom, packing her suitcase, handing her blond-haired daughter a hairbrush. In a box to the side, the magazine listed fifty-one trade names for thalidomide, obtained from Helen, under the headline, "Don't take these!" *Life's* thirteen-page spread examined the debate thalidomide engendered on abortion and euthanasia as the Vandeput trial unfolded and discussed how some doctors delayed respiration at birth or removed oxygen from a dying adult. Lawyers and doctors interviewed favored legal protections, while clergy denounced them. Parents with brightly smiling thalidomide babies told the magazine their children were gifts of God.

Helen expressed qualified support for euthanasia to the magazine. "In my opinion, euthanasia is the part of kindliness when no hope exists for life to be or to become of significance," she said. The difficulty lay in deciding when such a situation existed—for example, with an advanced cancer patient or progressively degenerative mental illness. The duty of the physician is to "relieve suffering and to enable the person to live a fuller life, but not to maintain life as long as possible. When there is no chance for improvement or for life to attain significance, it does not seem to me to be humanitarian."[30]

These were electric words, as the editors called them, and in the context of babies without arms or legs, Helen left herself vulnerable to attack. She quickly regretted taking on so sensitive a topic in the popular press and acknowledged that it weakened her cause. To contain the uproar, she asked other journalists to delete remarks on euthanasia from her statements.[31] The debate in England about euthanasia would not solve the problem of phocomelia babies, she had written to Lear at the *Saturday Review*. The problem was preventing malformations and, for thalidomide babies, finding "the best training possible to give these unfortunate children so that they can be self-supporting citizens."[32] Not for another decade would the country grapple with the issues raised by the comatose Karen Ann Quinlan, whose parents won permission in 1976 from the New Jersey Supreme Court to discontinue her ventilator.

Finkbine's plight made Helen's photos of maimed babies relevant. Americans were deluged with them in August 1962 and by stories in top magazines like *Newsweek* and *US News & World Report* that portrayed Finkbine and other women as victims of unscrupulous drug companies. Two days after the spread in *Life*, images of deformed babies, their eyes hidden by black bars to shield their identity, appeared on the cover of the *National Inquirer*. ABC's Howard K. Smith also broadcast the images on television.

The industry counterpunched, warning that government regulation would thwart the development of lifesaving drugs. The *New York Times* reported the drug company view that under the proposed federal regulations, insulin could not have been tested on a child in a diabetic seizure back in 1921, saving her life.[33] Helen dropped a note to the *Times* medical writer, predicting, correctly, that dying patients could have access to untested drugs under exceptions to the proposed rules. She redoubled her argument to test drugs on pregnant animals when, to her disappointment, Helen's lab experiments with thalidomide had failed to produce abnormal offspring. She reported the outcome in her *Scientific American* essay. Most of the rabbits had aborted, probably due to a high dose of the drug. (Beuren's dog also aborted.)

A consultation with Josef ("Joe") Warkany, a pioneering researcher on fetus and drugs at the University of Cincinnati, strengthened Helen's conviction that if thalidomide itself could not be linked to birth defects in the lab, drugs did cross the placenta and dangerous drugs might be uncovered by animal testing. In 1940 Warkany had shown that rats denied vitamin B2 at certain times during pregnancy gave birth to deformed offspring, indicating that compounds crossed the placenta and could affect normal development. He did not yet know how. He also raised multiple questions about testing drugs on pregnant animals, noting how results differed depending on species and, at very high doses, produced abnormalities in a long list of commonly used drugs.[34] Skeptical of the link between babies and thalidomide, he changed his mind when the drug was pulled from the market and births of malformed babies ended.

Helen now defended tests on pregnant animals by saying the relationship between drugs and birth defects was just beginning to be understood, and one day the mystery of how thalidomide worked would be unraveled. She was partly right. The thalidomide disaster revealed that animals react differently to drugs than humans do, a discovery that changed drug testing. Eventually after tests of multiple species, some of them produced maimed offspring. Thalidomide is still not understood.[35]

*

Against AMA members and what she estimated were thousands of "drug people" opposing the bill, Helen rounded up more than one hundred pediatricians. They believed it would save lives, not impede medical progress, as the AMA argued, she told Oren Harris, Democratic representative from Arkansas, sponsor of the drug regulation bill in the House.[36]

Drug companies lobbied Helen at the same time they lobbied Congress. Her desk was filled with literature and academic papers for and against drug testing. Scientists mailed her drawings of chemicals and shared their theories—in one case, on why pregnant rats were affected by folic acid but not thalidomide. From the stockpile in her office, Helen liberally dispensed samples of thalidomide to scientists in the hope that someone would uncover its inner workings.

*

Helen's public comments on euthanasia, a subject debated calmly only in philosophy classes, roiled some. Her admonitions to women against taking drugs in pregnancy sounded prim. Never an eloquent speaker, now she had lost her calm. Her traipsing after the media—pushing herself and her data on them—had a tinge of desperation. Twice in essays, she attributed a thalidomide drug to Ciba, a company that had never manufactured thalidomide itself. She apologized profusely, asked medical journals to correct the mistake, and probably escaped legal action because a company official was her friend. She did not apologize, however, for offending CEOs, the Catholic Church, and the medical establishment.

This was different from her campaign to find a surgeon for her blue babies or, when surgery took off, her admonitions against unnecessary operations. Helen had neither discovered the link between thalidomide and birth defects nor treated children maimed by it. She had attached herself to Lenz, expanded on his research, and appointed herself spokeswoman for the cause of drug testing on the unborn. She had become a lobbyist, one who was not above lending grotesque photos of babies to the *National Inquirer* and, some thought, an amateur one.

"It was not her finest moment," remembered Richard S. Ross, a friend and the long-time dean of medicine at Hopkins.[37] Fishbein, the long-time AMA executive, would express umbrage that Helen campaigned outside medical societies organized to speak for doctors. "This shouldn't happen again," he said.[38] To Charlotte, Helen seemed to make herself more important than Kelsey, when Kelsey had stopped the drug from entering American shores. Yet she understood Helen's obsession. Helen and those who worked with her suspected causes all the time, but they never seemed to be able to pin them down. They would open one door only to find a new door. "But here was something tangible," Charlotte recalled, "and she was pretty horrified."

*

The Kefauver-Harris amendment to the Food, Drug and Cosmetic Act as approved by a joint conference committee passed Congress on October 4, 1962. The bill required safe and effective drug testing, including animal tests ahead of human trials, informed consent, government control of investigators and drug trials, and strict record keeping. President Kennedy signed it into law on October 10.

By then, sixty US investigators were testing thalidomide. The number would grow to two hundred. Within the year, Lenz's slides of thalidomide babies appeared in the first of dozens of medical textbooks. Kelsey had begun writing new rules for drug trials. The new secretary of health, education, and welfare, Anthony J. Celebrezze, proposed rules for public comment without first soliciting drug company feedback, upsetting them. His deputy did ask Helen to review them, though.

She had moved on. Realizing the public could no longer count on medical journals or the lay press to warn them of a global health emergency, Helen targeted the World Health Organization as a replacement. "What can WHO do to prevent disasters in the future?" she asked WHO director general Marcolino G. Candau. In her three-page letter, she told him doctors were upset by the four-month lapse between thalidomide's withdrawal in Germany and its withdrawal in Canada and Japan.[39]

In Lisbon, at the International Pediatric Conference in September 1962, Helen convinced 150 doctors to petition the group's president to urge WHO to take the lead in communicating any ill effects of drugs and market withdrawals to health ministers worldwide. Afterward she flew to Mexico for another conference on the drug. "Too much travel," she wrote Charlotte, ". . . but thalidomide is too important to miss."[40]

*

Frances Oldham Kelsey is famous as the courageous civil servant who prevented the thalidomide disaster and the exemplar of an era in which Americans came to expect that the federal government should protect them from unscrupulous companies selling harmful products. In August 1962, at the height of battle with drug companies, President Kennedy presented Kelsey with the President's Award for Distinguished Federal Civilian Service. Her career at the FDA flourished. She oversaw new rules for clinical trials.

She could not have been a hero without Helen. In spring 1962, four months after thalidomide had been withdrawn in three countries because of its link to babies with malformations, Merrell was still pushing for FDA approval. Kelsey had run out of options; her last ploy was to force the company to change its label to include a warning about possible side effects of neuritis. Once they complied, she had no recourse. But on March 20, 1962, shortly before Helen's return to Baltimore with her notebooks full of data, Merrell suddenly let its application expire and, without telling Kelsey, asked testers to return leftover samples of the drug. As far as Kelsey knew, Merrell could resubmit its application at any time. Two events led to Merrell's decision: it was tipped off by Grünenthal that Helen was overseas making a sweep of doctors' offices to investigate thalidomide's link to maimed babies, and deformed babies linked to the drug appeared in a fourth country, this one on the American continent—Canada.

Merrell had delivered a letter to Helen two days after her return asking what she knew, hoping it still had a chance to market the drug. Helen kept the company guessing for a month, until the Distillers scientist successfully repeated his rabbit experiment and she shored up her facts. By then, drug companies had conceded thalidomide's problems and begun arming for a different kind of fight. When Merrell was finally ordered to contact doctors to get samples back, it could not; its record keeping was too sloppy. Just one of these tablets taken at the right time could result in the birth of a deformed infant. Worldwide, ten thousand babies were born with defects because of the drug. In the end, about seventeen thalidomide babies were born in the United States. This is because of Helen's suspicion that there was more of the drug out there than the company admitted, and her decision to personally raise an alarm in the absence of any federal or neutral party to protect peoples' health. Early, and often, beginning three months before the FDA publicly warned Americans, and at a time when Merrell recruited additional doctors to test the drug, she wrote letters, lectured, and called the press to warn of its dangers. Preventive health care worked.

She continued to advocate for warning labels on drugs. As Congress began debating how to combat dishonest advertising, she called attention to the ease with which manufacturers hid the same drug under different names around the world. In the *New England Journal of Medicine*, Helen revealed that drug companies in Europe had even added thalidomide to other compounds, to enhance sales, without any warning on the label.[41] Congress ultimately passed the 1966 Fair Packaging and Labeling Act, which forced companies to abide by truthful advertising for drugs and cosmetics. As the FDA budget ballooned, she urged Senator Humphrey to find Kelsey and her staff office space more respectable than a barn in a remote part of the city.[42] After a 1967 WHO meeting in Geneva, Helen urged Congress to ban the export of unsafe drugs.[43] Through the 1970s, she continued to suggest reforms and lobby for drug safety.

If Helen had stayed home instead of investigating thalidomide, Louis Lasagna might well have become the next FDA commissioner. The leading proponent of informed patient consent and the architect of controlled human drug trials—modern clinical pharmacology—he also believed the industry could regulate itself. The physician who got the job, James L. Goddard, a public health specialist, oversaw a 75 percent increase in drug recalls in his first year. More than four thousand drugs on the market were reviewed for effectiveness. The stakes were so high that the drug industry fought for the next decade, until 1973, when the US Supreme Court upheld the 1962 FDA amendments, allowing government to regulate drug safety.

*

At the start of Helen's career, colleagues asked why she bothered with children with heart defects. There was nothing to be done for them. In

Germany, people wondered why she had come to investigate. The drug was off the market. In the United States, scientists scratched their heads trying to figure out why she insisted on testing new drugs on animals when under the best of conditions, such tests would not have predicted the thalidomide tragedy. It was as if she had seen something they did not see.

No other doctor in a lifetime had examined as many children—thousands upon thousands—suffering from defects that left them struggling to live normal lives. Among her American critics, only Helen had comforted German babies born without arms and shortened legs. Once she realized that drugs crossed the placenta and could hurt the fetus, Helen saw what lay on the path ahead: a fire with the potential to consume the next generation of children. Who would be such a fool to let it burn? Not someone who had spent her life trying to ease the suffering of children. She grabbed her bucket of water.

She battled based on her own moral compass, purposefully tugging at peoples' emotions, anticipating that the science would follow one day, and even if a test was not foolproof, if it saved some babies from a life of suffering, it was worth the risk. Advocating a test of new drugs on pregnant animals when it might not predict all birth defects seemed to conflict with Helen's doctrine of frugality. By her calculation, the cost to society to care for infants maimed by drugs was greater than the cost of testing all new drugs. Unnecessary suffering, rather than unnecessary tests, was her denominator.

Her instinct was to reduce suffering, whether that meant tending to a houseguest who caught her hand in a door jam—she rushed over with a wet towel and massaged it—or helping a thalidomide baby an ocean away. The infant girl with tiny fingers hanging like stubs from her shoulders whom she had examined for less than an hour in London remained on her mind as she flew home to alert Americans to the inadequacies of drug testing. She could not shake her horror at the suggestion from the girl's doctor to amputate the fingers.

On her second day back at the office, between calls to federal officials and doctor groups to arrange meetings, Helen wrote to an orthopedist in Münster seeking his advice in the hope of preventing further mutilation of the infant. Helen asked the professor, Oskar Hepp, of Orthopädischen Universitätsklinik, to verify her belief that it was better for thalidomide babies to "use these little extremities as much as they could and their fingers should help them manipulate the prostheses?"[44]

"Never amputate anything," came his reply. Whether she passed his advice to the British doctor is unknown. She did use it to help others. Concurrently with her lobbying campaign, Helen later referred a German couple seeking to help their child to the best orthopedic doctor in Westphalia.

*

In January 1963, four months before her retirement and one year after she learned over dinner of a strange malformation in Germany that might be linked to a sleeping pill, Helen appeared on the cover of *Modern Medicine*.[45] The editor, Irvine H. Page of the Cleveland Clinic, sent her an advance copy, saying he liked her "cover girl" photo but, knowing it was hard to please everybody, what did she think? Helen's staff approved, she told him. In a V-neck print dress, a necklace of gold flowers, and black-framed glasses, the now gray-haired Helen wears the same dimpled half-smile she wore in 1959 as she walked in the Harvard procession to receive her honorary degree.

In the aftermath of her thalidomide research, she successfully solicited a grant from the March of Dimes to study the causes of birth defects. She suspected genetic mishaps more than environmental causes. She was in line to lead the American Heart Association. At home, too, she reaped the rewards of hard work. The winter bulbs she had planted in her greenhouse were a stunning success; by staggering her bulbs, she told Charlotte, she had hyacinths, daffodils, and tulips blooming at once. She and Amy positioned her two-year-old amaryllis in bloom again on the steps by the window.[46] When she returned home on a cold February night, she was surrounded by beauty she expected only in spring. Her home never looked prettier.

She expected to continue to host biannual meetings and barbecues, just as she would consult on patients and keep a hand in her clinic. Helen was pushing Neill as her successor, but a search had turned up a young doctor who had moved pediatric cardiology forward in an area that Helen was slow to embrace. The leading candidate, Richard D. Rowe at the Hospital for Sick Children in Toronto, had pioneered the use of left heart catheterization techniques on the newborn. Doctors like him were working to make it possible to operate on infants.

By April 1963, a second search was under way, this one for an artist. Neill and others wanted a permanent memorial to their beloved mentor, a portrait that would put Helen in the same league as the distinguished men who had developed and advanced American medicine and whose images regularly inspired and beguiled students and alumni.

Over filet mignon at the Hotel Belvedere in June 1963, Helen's colleagues toasted their mentor upon her retirement and lamented their double loss— Helen's clinical standards and her advocacy for children. The historical view of Helen was taking shape, and true to form, Helen would guide it.

24 PORTRAIT OF A PHYSICIAN

The doctors who had studied in Helen's clinic wanted a painting of their Dr. T to reflect her now-legendary status. She had changed medicine. A portrait of the regal blue-eyed Bostonian in a sufficiently weighty frame to hang in Hopkins's halls alongside those of the men was an appropriate gift to mark her achievements. Such paintings had been anticipated by Mary Elizabeth Garrett when she hired the nineteenth-century society painter John Singer Sargent to make portraits of the school's earliest doctors and then, of herself. Physicians in the late nineteenth and early twentieth centuries were among the most influential members of society, and portraits verified their status. It was a tradition.

Helen did not want a stuffy portrait. When her deputy, Catherine Neill, raised the subject, Helen let it be known that she wanted something simple, like the sketch of Park by the American artist Andrew Wyeth, known as the "Painter of the People." She admired the pen-and-ink drawing so much that she hung a copy in her living room. Wyeth had drawn it as a favor to Margaret Handy, the pediatrician to his children, to mark Park's eightieth birthday. Of no small consideration to Helen, Wyeth had also famously painted Handy. A copy of his 1949 portrait, *The Children's Doctor*, hung over the fireplace mantel in Helen's bedroom. The portrait was charming, a side view of a lovely, wise woman, her fingers on her forehead. In the background, he painted Handy in her overcoat, medical bag in hand, at the ready. It was Handy who had doubted the young Helen after stumbling upon her cow hearts in the boardinghouse bathtub. Of the two, Helen turned out to be the more accomplished.

Park broached the matter with Wyeth, who recommended his younger son, Jamie. A commissioned portrait was a rite of passage for an artist. Only sixteen years old, Jamie Wyeth had been painting since childhood, eight hours a day, under the tutorage of his father and his aunt. The committee members who met with him and viewed his work at Chadds Ford on a fine spring day in 1963 returned awestruck. Modernity had long since set in. Abstract expressionism was giving way to minimalism.

A contract was signed. Helen was pleased. The Wyeths were an iconic and self-made American family with roots in New England. Their dedication to their art reached back generations. Helen had grown up seeing N. C. Wyeth's illustrations in the *Saturday Evening Post*. Only Charlotte, who was on the portrait committee but who did not visit the Wyeth farm, questioned hiring someone so inexperienced for such an important portrait. When Charlotte voiced her concern privately, Helen dismissed it. The conditions of success were the same for scientist or artist. "Genius matures young," she wrote her friend.

*

Jamie Wyeth would study his subject in Cotuit, where Helen could put on her best face. She was a different person there—less tense, happier—though the painter's knock on the door of her cottage interrupted that ease. Now, in a sense, she was the patient. Wyeth followed her around, sketching her as she made sandwiches. She fiddled with her beads, pushed loose hairs into a net around her face. For two weeks she remained under his gaze. No infants struggling to breathe whom Wyeth might have seen react to Helen's calming voice. Wyeth, too, was under stress. He stared into his blank canvas, terrified. On endless beach walks Helen quizzed him about his method; they shared details on their doting fathers, how they honed their observation skills. She told him to do what he thought was right.[1]

That sweet grandmother image was not right, he knew; it was not real. Her intensity compelled him, as it had others. Again and again he returned to the contrast between what he expected and what he found. He dreamed about her determination in the operating room with Eileen Saxon. From the uneven way she powdered her face and the clothes that hung from her, he sensed her modesty. Finally, he had her sit on a chair and stare directly at him. This would be the sketch for his painting and the model for later portraits that would make him famous.

On the phone with Charlotte, Helen had been vague. "It's coming along," Helen responded when Charlotte inquired about it. "Don't ask," she said another time. Then, one day, in a letter inviting Charlotte to Cotuit for Labor Day, Helen added a postscript. "The painting is done," she wrote, "for better or worse."[2] The unveiling was scheduled for the following spring, to coincide with the hospital's seventy-fifth anniversary and a week of activities celebrating children's health care and the opening of a new Children's Medical and Surgical Center.

When Engle, the Cornell physician, gathered several committee members in her living room to preview the portrait a few weeks before the official unveiling, she was so shocked by what she saw that she called her children from their bedrooms to verify that it looked nothing like the doctor they loved. She wanted to ask Wyeth to change it. The committee chair, Buffalo pediatrician and art aficionado Ed Lambert, said they had no choice but to pay the artist and accept it. He considered it triumphant, a "monumental piece that would last through the ages."[3] Charlotte kept an open mind. Late on Tuesday afternoon, May 12, 1964, with more than one hundred of Helen's colleagues gathered in the Welch Library for a cocktail party, Charlotte and Lambert pulled the cloth off the painting.

To her last days, Charlotte would not forget the collective gasp she heard as the cloth fell to the ground. From the canvas, Helen's blue eyes gazed piercingly at her. Pieces of hair whirled around her face, out of place. A dark cloth draped around her chest exposed enough of her shoulder to be provocative. Her crinkled forehead and downturned jaw made for a frown. Eyes afire, lips in rebuke, hair flying, in a state of undress—Helen looked as if she had been flying through the wind in search of an enemy. The image that came to Charlotte's mind was that of a witch.

Charlotte burst into tears. Among the men, only Robert Cooke, chief of pediatrics, dared say he saw anything familiar. "Helen, you look as though the fluoroscope had broken down that day," he teased her.[4] Helen smiled, thanking people as they balanced sandwiches and glasses of sherry in what Charlotte realized later was a carefully planned response. She talked about the great honor of having her fellows present the portrait. Because of Helen, Charlotte would recall, the painting was but a stumble in an otherwise glorious day.

Blalock had hoped to be there and promised to try if he felt better, but he needed his energy for the keynote speech two nights later. Helen dissuaded him, telling him the portrait would be controversial. He had warmed her heart with his intention to be there, she wrote, but he need hardly tax himself for her.[5] Despite her sincere concern for his health—he would soon be diagnosed with cancer—her note read like a sigh of relief.

*

She and Blalock were a team despite everything. In all their skirmishes, they had long ago dropped any pretension. He called her Helen. She called him Al. Helen and Al. Neither one much cared for the other or felt comfortable with the other, but there was a respect, a fondness for days gone past and apprehension about the future. They had been nominated more than forty times by an international cast of scientists for a Nobel Prize in medicine and recommended several times by a Nobel committee. Their work opened a field of knowledge, but the prize had rarely been awarded to a surgeon.[6] She had seen less of him as they moved on different paths. His life had stabilized; he even remarried. His chief residents, now leading surgery departments around the country, banded together annually to celebrate his legacy in what they called the Old Hands Club. Helen's speech honoring Blalock at his sixtieth birthday at the Southern Hotel in 1961 had been a major success with them. Now, the Old Hands Club arranged for hospital trustees to surprise Blalock after his speech by announcing they would rename the clinical sciences building for him.

By then, Engle had secured enough Knights of Taussig in her corner to carry out her plan to permanently remove the unworthy portrait of Helen from Hopkins. Over lunch, she told Helen they were giving it to her. Helen agreed, saying she did not much like it, either. She also agreed with her friend Lambert that it was of high quality, enough to keep as an investment. To comfort her, Engle realized, Helen changed the subject, to her upcoming adventure as the first pediatrician and first woman president of the American Heart Association.[7] Helen brought the portrait home and stored it wrapped in paper in her attic, face to the wall. Except for a brief appearance in a midtown Manhattan art show, it remained in her possession for a quarter century. Helen happily lent the painting for Jamie Wyeth's first exhibition at the famed M. Knoedler & Co. Gallery in late 1966. She also suggested a title: *Portrait of a Physician*.[8]

*

Stunned to see a bevy of young doctors follow her aunt down a hospital hallway asking her advice during a visit in the 1950s, Helen's niece Polly realized the woman she had known as "very shy, awkward, just awkward" had blossomed. Helen's second blossoming occurred in the 1960s.

Some colleagues associated her "mellowed" demeanor with improved hearing. As she embarked on her role as elder stateswoman of medicine, Helen in December 1963 decided that a surgical procedure called a stapedectomy had matured enough to risk. It involved replacing a tiny bone in the ear with a metal one. (She was nervous enough to ask the eighty-six-year-old Park to accompany her for the procedure on her left ear. In hospital scrubs and seated in a chair next to her, he held her hand throughout.) An operation on her right ear followed a year later. Afterward, Helen's closest male fellow, Dan G. McNamara, never again saw "the look that Wyeth captured on HBT's face several times when I or someone else would foul up."[9] But Helen's improved hearing lasted only a decade, so it wasn't the only factor. She was no longer under siege within her workplace or stressed by daily life-and-death decisions for patients that depended on her own data gathering. She had earned renewed international respect for her role in thalidomide. She had successfully navigated from doctor surrounded by adoring patients to public intellectual and global advocate for women and children's health. Lyndon B. Johnson in September 1964 bestowed on Helen the Presidential Medal of Freedom, the nation's highest civilian award.

It was not without pain, this new platform. To Neill, the Helen of the "extraordinary" Wyeth portrait was "indomitably handsome, but aging and alone."[10] This description by a deputy of more than thirty years hinted at the struggle Helen confronted as she entered the last stage of her life. She missed her daily physical and emotional interactions with patients. Her advocacy on their behalf invited criticism. She needed friends and made time to cultivate them.

Her thalidomide campaign showed Helen that one need not be an economist like her father or a politician to influence law and public policy. She opened her two-year term as president of the American Heart Association in 1965 with an inaugural address that criticized the AMA for opposing national health insurance and international cooperation on drugs.[11] The heart association's own magazine, *Circulation*, refused to publish it. Undeterred, Helen once again put herself in the crosshairs of the influential AMA, this time at the height of its battle against expanding Social Security to provide preventive care for the poor and cover hospitalization and other medical care for the elderly. She was second only to Benjamin Spock on a list of prominent physician supporters. (She called the AMA's opposition to free care for the poor "extremely stupid." A system of private practice with some sort of government insurance was inevitable, she predicted.[12]) The lobbying was so fierce that in Maryland at least the AMA assessed member-doctors an extra $50 (about $422 today) to fund it. Congress passed both bills with overwhelming support. Helen was already in Cotuit when President Johnson invited her to join him and Harry S. Truman in Independence, Missouri, on July 30, 1965, where he signed into law amendments to the Social Security Act creating Medicare and Medicaid.

Helen's outspokenness irritated members who expected the heart association to focus on research and education. She upended that tradition too to promote prevention of heart disease over reducing adult mortality. Studies pointed to poor diet and little exercise, beginning in childhood, as the cause. She urged physicians and the federal government to emphasize improved diets and active lifestyles for children.[13] The first to highlight blood pressure as a problem in children, Helen also urged pediatric cardiologists to study its path to try to prevent atherosclerosis in adulthood. Recognizing that health care spending was exploding—it doubled as a percent of gross national product in Helen's lifetime and since then has nearly doubled again, to 18.9 percent—Helen tried to influence what it bought. Her enemy was preventable suffering, no matter the scale.

As early as 1964, she pushed for fetal research, including on fetuses of women who took drugs and planned to abort, in the hope it would provide for more treatments and cures and more healthy babies. As unthinkable as it then seemed, scientists would one day unlock genetic codes and begin to alter genetics to help people, she told a reporter.[14] "Once we find out what genetic forces produce abnormalities," she said, "we can then proceed to do something about them."

In 1967 after Christiaan Barnard announced the first human heart transplant, prompting dozens of copycat attempts, Helen called them out as an ego trip for surgeons. Heart-transplant patients died within days; if their bodies did not reject the foreign organ, they succumbed from infections contracted from the bacteria-laden pumps reused from trial operations in dog labs.[15] "Until we can figure it out, this remains a research procedure," she wrote.[16] Her public shaming intensified a few years later when, after $3 million and one hundred operations, still no transplant patient could live normally. At the time, the US infant mortality rate, a key indicator

of the nation's health, ranked sixth from the bottom among compara-
tively wealthy Organization for Economic Co-operation and Development
(OECD) countries. (It is now the third worst.) "If we can't afford to give
basic care to everyone," Helen wrote, "I don't see how we can afford to give
highly specialized care to a few."[17] Pressure by the American Heart Association led to a moratorium on the procedure except in Norman E. Shumway's
laboratory at Stanford University.

Realizing that the research medical school admired the world over for
its ability to provide specialized care and advanced treatments did not produce doctors needed for basic care, Helen in 1974 circulated a multipage
plan inside her alma mater for a new "school of health maintenance" that
could train what today are known as physicians' assistants. In this way
she spotlighted health care delivery problems that have created a distinctly
American conundrum: the United States spends twice as much per person
on health care, partly because of excessive use of medical technology, and
its citizens suffer from more chronic illness and die sooner than people in
comparable high-income countries.[18]

She continued to herald less invasive treatment. Open-heart surgery on
infants was well underway in 1966 when she congratulated her friend William J. Rashkind at the Children's Hospital of Philadelphia for his revolutionary method to use the catheter to create pathways in infants with
transposed arteries, eliminating the need for the palliative surgical procedure developed by Blalock and Hanlon (atrial septectomy) and delaying
reparative surgery until it was safer. "Too much operating is a mistake," she
wrote to her fellow McNamara after observing a new operation in which a
child died. She explained to colleagues what surgeons should have done
so routinely that people mistook Helen for one. No, she replied with her
trademark half-smile when a congressman asked if she had operated on
blue babies with Blalock. "But I am a very good parlor surgeon."[19]

Famous for saving children, Helen confounded some admirers with her
influential work to reform abortion laws. She advocated for mothers whose
suffering she witnessed—the life before her, not the life to be, or fetal life,
as she put it—and her belief that the decision to abort a fetus was between
a woman and her doctor. For Helen, any requirement for approval from a
committee was "an infringement on one of the most personal rights of a
woman."[20] Nor did she believe anyone else had a right to know about that
private decision. Angrily she rejected an invitation from Gloria Steinem

seeking prominent women who had abortions to out themselves in the inaugural issue of Steinem's *Ms. Magazine.* Helen was infuriated over their "militant" public strategy, she told Steinem's organization, since abortion was a private matter between doctor and patient. (Nor did she qualify for the list, she told the magazine.)[21]

A sought-after witness in the decade after thalidomide and Sherri Finkbine's flight to Sweden for an abortion, Helen frequently testified in Annapolis in support of abortion rights and in 1967 telegrammed lawmakers urging them to vote for House Bill 88. It passed, decriminalizing abortion in Maryland, a Catholic but liberal state, in one of the earliest abortion rights victories. For years until the US Supreme Court ruled abortion legal in *Roe v. Wade,* she continued to testify in Maryland and in Congress against efforts to restrict abortions based on the health of the mother. Angry letters followed. "The right-to-life people are my arch enemies," she told a reporter.[22]

Her focus on unnecessary suffering led Helen to act on seemingly disparate fronts. In Saigon before the United States bombed North Vietnam and America erupted in antiwar protests, Helen visited orphanages to call attention to war's injured and abandoned children. After learning that a neighbor's son survived a lightning strike despite turning blue and being taken for dead, Helen investigated the impact of a strike on the body and explained to lay readers and medical professionals alike how prolonged resuscitation could save a victim.[23] She published her own shortcuts and coping strategies gleaned over a lifetime to help others manage hearing loss.[24] And in a 1973 note that galvanized male Hopkins administrators, she pointed out that the doors into the Park Building, designed to keep in the heat, were so heavy that a mother carrying an infant could not open them without fear of dropping her infant! Panic ensued when the replacement cost came in at $4,000. (The fire department approved a $1,900 hold-open device.) On her own initiative, Helen pushed for international exchanges on drugs and health issues through the fledgling World Health Organization and participated in professional exchanges abroad. Dutifully she sat on committees, but the work bored her; as she told colleagues, she no longer had the luxury of turning off her hearing aids. Helen countenanced suffering by dogs over that of her patients, regularly making impassioned arguments to Congress in defense of animal experiments. Not until 1966 after *Life* photographed a Maryland farm with more than one hundred chained, filthy, and skeletal-like dogs in cramped quarters, waiting to be sold to the state's two research medical institutions, did Helen publicly concede animal care regulation might be needed.[25]

She was happiest sharing her expertise. Hosted by medical associations and former fellows who had opened children's heart departments of their own, Helen lectured, taught, and examined patients in great cities and small villages in Latin America, Europe, and Asia in the 1960s and 1970s. For several years she taught for two months at a time in India, including at the Christian Medical College and Hospital in Vellore.[26]

Radiating joy and wearing her trademark pearls, Helen appeared so elegant and adventurous sitting atop a camel in Jerusalem that her sister Catherine advanced it for her official portrait. As a symbol, it was not far off. In what for women amounted to a desert, Helen kept riding long past the point where others dropped to the ground. The advent of jet engines helped; instead of crossing the ocean on the *Queen Mary* for weeks, Helen was deposited at her doorstep in hours. "Isn't flying wonderful?"[27] she wrote to Charlotte.

*

Helen wrote one-third of her papers in her emeritus years. She went to sleep exhausted and hoped the phone would not ring. "Distinguished Woman Doctor Still Works a Full Day at 73," read the headline in the *New York Times* in spring 1972. A Harvard undergraduate anonymously congratulated Helen for encouraging her when her father, a Harvard surgeon, regularly dismissed her desire to become a doctor, saying "women are just physically incapable of enduring the long hours and hard work of my profession."[28]

In Boston, Helen was now a celebrity. One evening in May 1973, an elderly woman arrived in distress with an irregular heartbeat (atrial fibrillation) at the Harvard-affiliated Mount Auburn Hospital in Cambridge. She told the admitting resident on call that she already had taken an extra dose of digitalis, a heart medication, on advice of her sister. The surprised young resident, Heinrich Taegtmeyer, soon found himself on the phone with Helen, whom he knew as one of medicine's icons. When Helen arrived from Baltimore the next day to oversee her sister Mary's care, she included him on her team.

While word of Helen's presence spread, Taegtmeyer and other residents presented an interesting patient to Helen: a young man in his twenties with an infected hole in his heart (ventricular septal defect). Helen demonstrated her technique of palpation before she took a special stethoscope from her purse. Then she lectured the small group of students and young doctors at the bedside. The grateful residents convinced her to stay for a few more days to give them her famous lecture on blue babies.

What followed was the tale of three thousand children who underwent successful heart surgery. Helen had just completed a major study that cataloged not only their health status, as ordinary follow-up studies would report, but what they had done with their lives. Now she could boast about them. Of original blue babies, for whom the artificial bypass to the lungs was a palliative operation, 50 percent were still living almost twenty-eight years later. Among those who had a second operation to reconstruct the heart using the open-heart method, 79 percent were still alive. Of 319 children born to these patients, six had heart defects. Of those contacted, 35 percent graduated from college. Eight percent had doctorates. Some were CEOs, 20 percent were professional or semiprofessional, and most (28 percent) were clerical, sales, and technical workers.[29]

Of the first 685 blue babies (1945–1951), many of whom came from families that could afford to pay for their care, 53 became professionals in law, medicine, religion, education; 32 were small business owners or technicians. Another 80 were clerks or skilled laborers. Only 2 percent were disabled and could not work.[30] Her goal in describing how people flourished and supported themselves was to demonstrate that a handicap in early years was not an impediment to success in adult life. On the contrary, she wrote, "the handicap motivates them to do their best."[31]

It was amid gathering data for these follow-up studies that Helen's office assistants tired of shouting and repeating themselves. In response, at the age of seventy-five, Helen outfitted herself with modern hearing aids and began what would be sixty-three sessions over four years with a private tutor, Rose Strauss, to learn and practice strategies that enabled her to communicate for the rest of her life.

Within days of her impromptu lecture, Helen learned she had been elected to the National Academy of Sciences. Membership in the Academy was her most coveted award. Her election followed Blalock's by almost three decades. The Academy in the interim had added a branch for medicine. How long a campaign was waged to establish Helen as a scientist, and who waged it, among a membership of two thousand, mostly men, who conducted science in the laboratory, is an Academy secret. At a time when two-thirds of Americans still believed that society would be better off if a man became professional and the woman managed the home and when young women battled the widespread perception they would never be equal to men and should not even try, she represented a woman who had succeeded in doing just that.[32] When the fifteen other female members of the Academy asked Helen to help expand the pool of women scientists, she said yes.

"Many a time I felt that I was unwanted which was in a sense true! Of course, even today men get much better jobs and are promoted ahead of me. I'm worth as much to the university as any of them & I get a great deal less. But I feel that in the background," she wrote to Charlotte in 1953. "I enjoy my work and fundamentally I've always worked for the pleasure of working. There I'm fortunate."[33] When a few years later, she was promoted to full professor, her response was nonchalant. Her appointment six years before her retirement along with a host of others came during a regime change. It followed by two years that of a male colleague in her own department.[34]

It is the work that counts. Her fight for recognition nonetheless distinguished Helen from many women who remained in the background. She mentioned slights she had endured because of her gender, often with backhanded humor. Speaking at the unveiling of a new technology, an oxygen therapy machine, for example, she thanked Harvard for not admitting her because she would never have had the opportunity to collaborate with Blalock. Accepting the AMA's top award—apparently she made up with the AMA—she noted her partner Blalock had been similarly honored decades earlier. She told her story without rancor. This was just the way it was, she would say. For pointing out inequalities, she was criticized by male colleagues for carrying a chip on her shoulder.

*

Park in his last years finally understood why Helen was so adamant about putting her name first on the blue baby surgery paper. From Helen's written account of her work, he realized how early she understood the solution and how often her pleas for help "fell on blind eyes." She suffered as Columbus did, he wrote, from seeing the possibilities so clearly but being unable to convince any sponsor.[35]

A quarter century after he photographed Blalock, in May 1975, Yousuf Karsh returned to Baltimore to photograph Helen. He learned that "one could not have existed without the other" when he proposed an exhibition of photographs of physicians and scientists he had taken on contract, called "Healers of Our Age," to mark Harvard Medical School's seventy-fifth anniversary.[36] Mellinkoff, dean of medicine at UCLA, who had rotated through Helen's clinic, noticed Helen's omission and helped arrange her portrait.

On the eve of making her portrait, Karsh and his wife, Estrillita, dined with Helen. As her dogs roamed underfoot, the talk at the table drifted to Cotuit, her futile effort to enroll in Harvard, and how women doctors were still paid less than men and viewed as inferior. But the most memorable topic was the healing power of touch throughout history. Her guests had noticed Helen was using hearing aids and reading their lips. Helen was grateful, she told the Karshes, not for the deafness that forced her to listen to the heart with her hands, but because of what it gave her, a greatly enhanced relationship with patients.[37]

Famously, Karsh reverently photographed the hands of many of his subjects, including the great doctor Albert Schweitzer, the heart surgeon Michael DeBakey, and Helen Keller, who inspired the world learning to read braille with her hands.[38] That evening, Karsh recounted how the blind and deaf Keller had put her fingers on his face, by way of greeting, and photographed him with her fingertips. But the photo he took the next day that would become the most loved image of Helen is of a slightly bent white-haired grandmother figure cuddling a baby, and gently pressing the infant's chest with a stethoscope. It would be printed on programs, in books, and framed in offices and hallways. Estrillita included it in a permanent display of her husband's work in Brigham and Women's Hospital in Boston. That surprised doctors like pediatric kidney specialist Robert I. Levy who knew Helen as a unique master of the rarely used technique of listening to the heart with her hands. The Helen who grabbed his stethoscope and admonished him with words that stuck with him for five decades—"the hand before the ear"—was the one in the Wyeth portrait.

As undefinable as she sometimes seemed to colleagues, she was even more elusive in oil. Helen's fellows presented her with a second portrait in May 1970, but it was quickly dismissed. There was something odd about the arm. A third portrait, commissioned by Engle in 1982 and painted from a photograph because Helen refused to sit for it, shows a white-haired woman in pearls and a doctor's coat. It was not very good either, but Helen's fellows agreed it should be the one to hang on state occasions.

She was not demonstrative, but she was always cheery, usually laughing. In Cotuit, Helen's nieces and nephews were transfixed by the international reach of her friendships. Adults became fully immersed in her afternoon adventures crabbing, hunting blackberries, and making jam. Former fellows familiar with her rigorous professional methods were simultaneously stunned and delighted by her rituals in leisure; pediatrician Tony Perlman, who had helped clear Helen's hillside thirty years earlier, and his wife, Patsy, learned not to clear the soup bowls before Helen served lobster from a rolling table at her side. She had boiled each in its own pot for two minutes and wrapped it in newspaper to finish cooking just as they'd completed the first course.

Those unbothered by her precision found her endlessly fascinating. Estrellita Karsh came away from her three-day visit with Helen feeling a chemistry like that of falling in love, as when "you want to keep on knowing a person." Seeing Helen wander her grounds in the company of her Irish setter and energetically toil in her garden, Helen's boarder in later years, Mark Eaton, a Hopkins undergraduate, was reminded of another elegant woman with the same high cheekbones, also from New England, also private and firm in her opinions, similarly athletic but three inches shorter—Katharine Hepburn.[39]

In 1975, the year of the Karsh portrait and Helen's definitive interview in Cotuit by Boston Children's Hospital chief Charles A. Janeway, Hopkins medical school was still not a friendly place for women. They were paid less than men. They lacked mentors to promote them. One of those who arrived that year, Janice E. Clemens, a microbiologist and geneticist, would fight to succeed her mentor, who had not recommended her. She stayed partly because she found another female researcher who, like her, was married and raising children. It would be decades before she could display family photos on her desk.

As a new generation of women doctors settled in, Helen faced the ramifications of a system where title and salary had been bestowed infrequently and rarely to women. In the mid-1970s as she contemplated moving to a Quaker retirement community in Kennett Square, Pennsylvania, Helen began renting her Cotuit cottage in August, the most desirable vacation month, to cover expenses. She declined invitations to lecture or attend conferences, telling organizers she could not afford to come unless the host paid her travel expenses. Helen's inherited wealth suffered when an investment soured. Embarrassed that someone of Helen's stature should need to ask for travel funds and angered that Helen's pension, like her salary, was less than that paid to men of her era, Charlotte and two other former fellows wrote to the medical school dean, asking him to improve Helen's compensation. Helen got lifetime income in exchange for giving Hopkins $150,000, possibly from the sale of her house. Details remain private, but Helen was satisfied.[40]

To raise money, Helen sold her Cotuit cottage to Hopkins at a price below market, in a deal that allowed her to use it during her lifetime, with the expressed hope that the place where she renewed her spirit would be enjoined for use by Hopkins faculty. From the grave, she wanted to remind colleagues to heal themselves.[41]

Helen's relationship with money, like her demand for credit, was rooted in justice. Nothing in his lifetime left Blalock as incensed as Helen's 1953 request that he share a patient's $15 donation; hadn't he paid for the catheterization lab and other tools that benefited her patients? For Helen, who had been alerted to the gift by the donor, it was a fairness issue. Pediatrics could not hope to attract the donations grateful surgery patients poured into Blalock's department for work she helped make possible.

While she pursued small sums she felt due, she gave away large amounts of her own money. Anonymously she underwrote promise in others who because of poverty or birth faced an uphill climb to reach their potential. There was the college tuition for a gifted but impoverished music student. There were down payments she made that allowed Black men to obtain mortgages and buy homes on Maryland's Eastern Shore.[42] Her papers reveal gifts of more than $100,000 in her lifetime ($773,000 today) to help educate and employ Black men and women, including to build a sewing factory. The builders sat in her living room. From a letter she wrote seeking additional donations, the project may not have come to fruition.[43] What other seeds did she plant with this gift?

She plotted her departure from the house she built in Baltimore almost as carefully as she drafted national standards for doctors who succeeded her in pediatric cardiology. David Alston, a young brand designer, did not realize he was being vetted when she invited him and his wife to her house one Sunday and her neighbors showed up. His was the lowest bid but Helen picked him anyway. She stayed six months through spring 1977 to oversee the transition (the wait required for a first-floor apartment that would allow her to sleep outside). Saturday mornings, she convinced Alston to work beside her in her informal beds. "Let's go dig up the daffodils," she pushed him. Dig, divide, transplant. The younger man felt he was being worked like a mule. "She's still bobbing along," he remembered thinking, "and I'm beat to death."[44]

Helen kept her practical blue Volvo when she moved in case she had to transport a sick patient to the hospital, although she envied the flashy red convertible her classmate Helen Pittman chose in retirement. Well into her eighties, she kept her hands nimble, sewing her own slipcovers and kneading bread, still hoping to unlock the cause of her patients' suffering.

At eighty-four she published her investigation into the genetic component worldwide of birth defects.

Certain that birds held clues to problems in the human heart, she also began to dissect warblers at her kitchen sink, using her fingers to pull apart heart muscle just as she had studied beef hearts many years earlier in her boardinghouse bathtub. Initially she asked friends to bring her carcasses they found in woods or under electric wires. She cut them open in search of malformed hearts. Ultimately, over three years, she examined five thousand hearts of birds in a laboratory at the Delaware Museum of Natural History. Her compilation of research on malformed bird hearts and thesis offering an evolutionary explanation was the most comprehensive at that time. While not scientific proof, it indicated to Helen that neither parents nor toxins could be blamed for most heart malformations.[45]

Evidently she worked in pain. In early May 1986, Helen returned to the city she had made her home as she did every spring to visit her gardens and her friends, staying several nights with Charlotte. One evening, late for a dinner in her honor by the Association of American Physicians in Washington, DC, she explained to Ross, the medical school dean, that she had been unable to fasten the clasp of her necklace. He did it for her and, later, after escorting her back to her hotel room, unfastened it, at which time she also asked if he would give her dress "a little zip down the back?"[46] Given the reach of Helen's long arms, one imagines she suffered from arthritis. She died two weeks later, in a hospital, following an automobile accident on her way home from voting in a primary election in West Chester, Pennsylvania. She was a few days shy of her eighty-eighth birthday.

Helen's fellows published her paper on bird hearts. Other ends were not so neatly tied up. Her father's silver tea service, which Helen tearily gifted to the dean of medicine, ended up on a dusty shelf. Her beloved Cotuit cottage was sold. And still inside her closet was the Wyeth portrait.

*

Helen's friends were determined that her accomplishments be remembered; the question was how. Would she be seen as the grandmotherly doctor, or the pioneering pediatrician, or the political activist? One thing they were sure of: they didn't want her remembered as the wild witch in Wyeth's portrait, which they thought they had securely hidden from Helen's alma mater.

Their troubles began when Helen's nephew, Gerard Henderson, donated the Wyeth portrait back to Hopkins, earning a tax write-off rather than incur capital gains. By 1989 the question of what to do with it had captured the art world. The new chair of pediatrics, Langford Kidd, who admired Helen but knew her only briefly, favored hanging it in the children's heart clinic. In *American Medical News*, the AMA's publication, he glibly summed up others' views of Helen as either a "saint on earth or a holy terror."[47] An alarm went out among some of Helen's fellows. For a quarter century they had kept the portrait out of sight, with Helen's agreement and expectation that it would never hang at Hopkins. They had nearly forgotten it. Suddenly their plan was imperiled. Old nemeses like Richard J. Bing advocated showing the portrait. An art news magazine took on the question of artistic freedom. More people wanted to see what Charlotte called "that dreadful painting."

An unflattering portrait was not the only thing that alarmed Charlotte. A draft of a book on the history of cardiology that Bing had asked Charlotte to read placed credit for the surgery solely with Blalock.[48] The camps that flourished during Helen's lifetime remained active. By 1991, with Blalock dead and Helen famous as the "patron saint" of pediatric cardiology, his admirers worried that her role in the development of the blue baby surgery had "in the minds of some, completely overshadowed Blalock's contribution."[49] (Helen, in her lifetime, even worried about this. "Be sure to give Dr. Blalock great credit" for developing the shunt in his laboratory, she told one interviewer.)[50]

Helen's fellows had to be careful. It would be unseemly for academics to suppress artistic expression. But women doctors including Mary Jane Luke, a retired assistant professor of pediatrics, were seething. Not a single building on campus bore the name of a distinguished woman doctor, she wrote Charlotte. How could Hopkins display such an unfortunate image of one of its most famous woman graduates, especially one that the women felt did not ring true? "It is distressing to think that the shocking visage of a Dr. T unseen by any of us should be forced to glare down from any wall at JHH," she wrote.[51]

She proposed to buy it back. Negotiations over the Wyeth portrait opened on Baltimore's Mount Vernon Square, in a private women's club across the green from where once stood the mansion of Mary Elizabeth Garrett. Over lunch at the Mount Vernon Club, only steps away from where ninety-seven years earlier, Garrett and friends had secured spots for women in medicine and provided them financial and emotional support, Luke met with Ross, the dean of medicine, in an attempt to guard the reputation of one alumna whose work fired the world.

The portrait was not for sale. But Ross sought consensus on what to do with it. Polls of Helen's fellows followed, permeated with entreaties from prominent Baltimoreans and cardiologists outside Hopkins whom Charlotte solicited to stack the deck against showing the portrait at Hopkins or anywhere with her name on it. A suggestion by Helen's closest male fellow, McNamara, to show it at the 1991 medical school reunion along with two other images met with swift condemnation. The Wyeth portrait was an insult to Helen, Engle told Ross. It made her into a "gaunt, evil person, totally unlike the HT that we knew and loved," she said. Charlotte offered a compromise that became the solution. If the portrait stayed at Hopkins, she asked that it be "fully" hidden from view until she and other fellows with strong personal feelings about it were dead. That might be in twenty-five to thirty years. "My opinion on this will never change," she told Ross.[52]

*

The depth of the reaction continued to surprise Jamie Wyeth. Few had seen it or even knew it existed, but Helen's portrait haunted him over the years as he would learn it had haunted Helen's friends. As he followed reports on the painting over the decades, he realized that it had mortified Helen's friends. This was the nature of portraits. Nobody liked them. It reminded Wyeth of portraitist Sargent's famous insight that the very definition of a portrait is "there's something wrong with the mouth." He thought of Picasso working years on a portrait of Gertrude Stein in Paris, and the day he finally brought it to her and unveiled it. Her friends too drew back in horror. But it doesn't look like her, they said. And Picasso replied, "It will."

That was what many feared; that people would remember Helen for her controversial portrait, not for her accomplishments. That like Picasso and Gertrude Stein, the artist would be more famous than his subject. Their concerns seemed justified when in 2008, with only Charlotte remaining to guard against a showing of the portrait in the place where Helen won her fame, Charlotte discovered that Helen's critics were "still trying to destroy her," as she put it.

"She was no scientist," Bing said about Helen in one historical account, "and complicated scientific matters disturbed her."[53] He credited Blalock squarely for the blue baby operation.

Making the portrait public would add to Helen's destruction. The belief that successful women doctors had been systematically labeled witches and burned at the stake by kings and popes, established by a 1930s-era academic paper, has been discredited.[54] But the idea of women doctors as witches remained embedded in the public consciousness. It was a way to dismiss women's accomplishments.

In the fifty years that Helen's fellows hid the Wyeth portrait, buying time for her reputation to solidify, Helen began to look less and less like a witch. To Janice E. Clemens who by 1999 had become the first woman vice dean of the Hopkins medical faculty, the Wyeth portrait resonated in a way that the stately image of Florence Sabin in velvet striped academic robes and Harriet Guild, Helen's contemporary, seated in doctor's coat, a stethoscope in hand, did not. In anticipation of taking over one of medicine's top jobs, she searched the secure repository maintained by medical school archivists for images of women doctors. As soon as she saw Wyeth's portrait of Helen, she chose it. To succeed, a woman had to be determined and serious to be taken seriously. "One look at Helen's eyes and you know that's a woman who is determined," she said.[55] This was before widespread use of the internet, which today hosts multiple competing images and essays about Helen, and when gender inequity was extreme (for instance, female anesthesiologists at Hopkins earned 14 percent less than their male counterparts.)[56]

*

The embrace of the Wyeth image of Helen by a powerful woman doctor at Hopkins was the first crack in the fortress Helen's friends had erected around it, though the copy she ordered hung only in her private office, and Helen's fellows were not consulted. Next, a male fellow studying under

Catherine Neill won permission to show a copy of the portrait along with other images of Helen during an out-of-town lecture introducing her to a new generation of doctors. Edward B. Clark, a military veteran who met Helen in later years, believed that only the Wyeth painting captured her "commitment, perseverance and power."[57] To Clark, it was the look that "allowed Helen to have leverage within a male-dominated organization."[58] (Ross argued to Helen's fellows that failing to show the portrait was like ignoring "legitimately derived data" in a study just because it did not fit with the bulk of the findings. Showing it would also put to rest rumors that the painting had been destroyed.)[59]

As the fiftieth anniversary of his portrait of Helen approached, Jamie Wyeth asked to borrow it for a retrospective of his work at the Boston Museum of Fine Arts. For five months in 2014, Helen's portrait hung in the exhibition gallery, next to one Jamie Wyeth made of his father, Andrew. Helen's portrait drew strong reactions from those who learned about the controversy in a *New York Times* article posted on the museum's website.[60] They praised the painter, subject, and the portrait's intensity. "A mindful individual . . . An intense, intelligent, caring woman . . . like a real, multi-faceted person . . . dead serious . . . awesome . . . beautiful . . . remarkable," the "doctor's keen intelligence comes through in her eyes." The artist "captures her heart and soul." Alluding to Sargent's definition of a portrait, a relative of Helen's suggested it would have worked with a little more coverage around the neck and a slight smile.

*

Helen's tombstone in Mount Auburn Cemetery in Cambridge is bare but for her name and the dates she lived. This is consistent with her belief that she lived not to be recognized but to serve others. On the whole, Helen told a reporter, she had done more good than harm. She lamented that people only read about the successful operations, not the sorrow and hard work behind them.[61]

She was a woman in a man's world and deaf in a job where saving a child's life depended on her detecting the slightest sound. Like the broken hearts she examined that found a way around their misshapen or missing parts to push blood to the rest of the body, she forged a unique path to fulfill her mission. She was first to diagnose heart defects in living children and to suggest how to fix them. Her solution evolved from observing how

Helen in a section of the gardens she cultivated on her land over a quarter century, with her golden retriever, Heidi, and her dachshund, Kleine Knabe (Ger. baby boy), 1973. Courtesy of the Alan Mason Chesney Medical Archives of the Johns Hopkins Medical Institutions.

the body naturally compensated and from listening to her patients and their bodies. What could not be proved in an animal laboratory, she proved in her laboratory of children and heart sounds and autopsies.

Helen's methods have been supplanted, though the Blalock-Taussig shunt outlasted attempts to improve upon it and is still used occasionally on fragile infants. And for certain conditions, physicians still extoll examining patients with the hand.

"In seeking truth, take one step at a time," she had said in her long-ago talk at Goucher College to women scientists. "If you do what you are sure is right, you will not be led astray and as the years go by you will contribute to the advancement of knowledge and the improvement of the welfare of mankind."[62] And this is how she became powerful.

The building that bears Helen's name has no physical presence. It is an ever-growing edifice of knowledge. Helen lives on in its own shadow, in the improved health of children, and in the generations of doctors and families she inspired. Her work is not done. On a walk around Lake Roland in early April, you can glimpse some of what she pursued. The slope from her house to the lake is dotted with more daffodils than any other, and above, hidden to the eye, her lawn boasts thousands of them, now in seventy varieties. This is the house that Helen built.

*

It is with pleasure . . . that I recall your helping hand and so must the countless number of children who down through the years received as I did a new and renewed lease on life and all that it has to hold.
—letter to Helen from a former patient, who had become the mother of two sons, one a law student at a California university, 1972

Although we were poor people, you never hesitated to administer to [. . .] and comfort us. [. . .] lived 32 years gently and patiently enriching the lives of all who knew her, because God in his infinite wisdom directed her to your attention.
—letter to Helen from the father of a young child Helen treated during World War II, finding her a bed in the overcrowded hospital by converting a utility closet, 1971

Congrats on your whole life, for being the warm, humane person you are.
—letter to Helen from the parents of a West Virginia girl whom Helen treated and who later died, 1975

ACKNOWLEDGMENTS

My debt is greatest to Barbara Morrison and Richard C. Reynolds, MD. Morrison, poet and teacher, believed in Helen before either of us knew the arc of this story and helped shape it at the expense of her own writing. Reynolds, who at the height of his career influenced hundreds of thousands of students and healed colleague and patient alike, in his twilight, his eyesight faded, read this work and enhanced it with memories of Helen, medical school, and good doctoring.

Charlotte Ferencz, MD, trusted me with Helen's story. I miss her. By the time of her death she no longer cared about the Wyeth painting of Helen. She knew she had helped make a different portrait.

Michael D. Freed, MD, cardiologist at Boston Children's Hospital, was adviser and critic; there is no way to adequately thank him. I am grateful too for the support and enlightenment of John W. Schatz, MD, FACC. Any technical errors are mine alone.

Jamie Wyeth generously shared his creative process, insight, and good cheer.

Mary Taussig Henderson, granddaughter of Helen's sister Mary, introduced me to the Cotuit skiff and family ways in summer 2012. Thank you, Mary, for being my faithful liaison. I also thank Mary's brother and sister-in-law, George "Bunker" and Dita Henderson, for their stories and hospitality in Cotuit and Boston, and Helen's niece, Polly Henderson Horn.

It was my good fortune to meet the late historian James W. Gould, who transcribed and alerted me to the diaries of James Herbert Morse, the Taussigs' summer neighbor, that document forty years of life in Cotuit. Kenneth

H. Molloy and Cindy Nickerson of the Cotuit/Santuit Historical Society, and Morse descendant William Beebee helped me access and verify the journals. David Churbuck described the geography of the Cotuit shoreline in Helen's time and the art of digging for clams.

Andre and Catherine M. Papantonio, Arthur and Shirley Ferguson, and Harry and Sarah Lord were my guides on a tour of Helen's Baltimore home and gardens.

Research and writing this book spanned a decade, fitted around work and family. Many who aided in its creation did not live to see its completion. My consolation is in knowing how they relished sharing their memories. They include the extraordinary Richard S. Ross, pioneering cardiologist and medical school dean who received me in his home despite his own suffering; the surgeon Jacob Haller Jr.; and pediatricians Tony Perlman and Kitty Esterly.

I owe much to the worldly Baltimorean Patsy Oppenheimer Perlman and many doctors who passed through Helen's clinic on their way to stellar careers, including Frank D. Milligan, Norman J. Sissman, Robert I. Levy, Henry A. Kane, and others who prefer to remain anonymous. Edward B. Clark, William N. Evans, Gerri L. Goodman, and Heinrich Taegtmeyer assisted me beyond their published articles.

No historical account would be possible without archivists and volunteers who collect, preserve, and make accessible documents that help reveal truth. This work depended on ten such collections. My debt is eternal to Nancy McCall, director of the Chesney Archives, Johns Hopkins Medicine, Nursing and Public Health, Baltimore, and her deputy, Marjorie Winslow Kehoe, whose scholarly and steady hand guided me through the papers that are the foundation of this book. Phoebe Evans Letocha, Timothy Wisniewski, Andrew Harrison, and Kate Ugarte also worked hard to accommodate me. I caused them "no end of trouble," as Helen said of difficult cases. I am also indebted to Rebecca Williams of the Duke University Medical Center Library and Archives; John Rees of the National Library of Medicine's history of medicine division; and John P. Swann of the US Food and Drug Administration.

In Los Angeles, Paul and Margaret Rosenthal opened boxes of family papers to me, as did Annick (Maurin) and Luke Zander, in Carteret, France. I am grateful for my dear friend and translator, Annette Cripps. Others who eased my way include Rosenthal family friend George Fritz; the incomparable

Baltimore Sun librarian Paul McCardell; Charlotte's niece, Clara Arnold; and Sarah Harris, editor of the *Daily Californian*, University of California, Berkeley.

From the start I benefited from the wisdom and support of writing friends John Bainbridge, Stephanie Citron, Rita Costa-Gomes, Laura Mason, and Barbara Morrison. It was Morrison who brought Helen to the attention of Laura Keeler at MIT Press. Keeler made this a better book, the mark of a great editor. I also thank the MIT Press team, including Virginia Crossman and Beverly Miller, Katie Helke, Suraiya Jetha, Marge Encomienda, Sean Reilly, and Katie Lewis.

At my side throughout has been my husband, Gary Goldberg, who grew to know Helen almost as well as I did. My gratitude is boundless. I also thank my best friend and sister, Susan DePalma, for her stalwart support.

Above all, I thank Helen for leaving the clues, the cases, and evidence she used to serve her patients. What a thrill it was to hear her voice as she unraveled mysteries of the heart and to learn from her own hand of her sorrow, frustration, and triumph.

NOTES

CHAPTER 1

1. Former patient to HBT, December 28, 1964, patient correspondence, box 33, Helen B. Taussig Collection, Chesney Archives—Johns Hopkins Medicine, Nursing and Public Health, Baltimore, MD. Hereafter referred to as the Taussig papers.

CHAPTER 2

1. Mary Taussig Henderson, Memorial Biography of Edith Guild Taussig (1942), Radcliffe College Alumnae Association Records (RG IX, series 8, 71v), Radcliffe College Archives, Schlesinger Library, Radcliffe Institute, Harvard University.

2. Robert D. Richardson Jr., *Emerson: The Mind on Fire* (Berkeley: University of California Press, 1995), 263.

3. Mary "Polly" Henderson Horn, daughter of Helen's sister, Mary Taussig Henderson, interview by the author, July 23, 2012.

4. Redvers Opie, "Frank William Taussig (1859–1940)," *Economic Journal* 51 (June–September 1941): 359.

5. Joseph A. Schumpeter et al., "Frank William Taussig," *Quarterly Journal of Economics* 55, no. 3 (May 1941): 350.

6. W. Proctor Harvey, "A Conversation with Helen Taussig," *Medical Times* 106, no. 11 (November 1978): 29.

7. In the nineteenth century, one in four people died of TB. Today the World Health Organization estimates a quarter of the global population is infected with the disease, and up to 5 percent develop it in their lifetimes. World Health Organization, *Global Tuberculosis Report* (2020).

8. Nicole Loraine Smith, "The Problem of Excess Female Tuberculosis in Western Massachusetts, 1850–1910" (master's thesis, University of Massachusetts Amherst, May 2008), 17, https://scholarworks.umass.edu/cgi/viewcontent.cgi?article=1179&context=theses.

9. Horn, interview by the author.

10. *James Herbert Morse Journals* 16 (1908–1909), September 30, 1909, 51. Historical Society of Santuit and Cotuit, Cotuit, MA. Morse, a Harvard-trained New York City private-school headmaster, spent his summers in a hammock, pen and notebook in hand, recording life in Cotuit.

11. Opened by Edward Livingston Trudeau in 1885 and named for him after he died in 1915, the sanatorium was a training ground for doctors as well as an innovative care and rest center for patients.

12. Cotuit Mosquito Yacht Club records, Cotuit Library, Cotuit, MA.

13. Radcliffe College Student Files (RG XXI, series 1, box 118), n.d., Radcliffe College Archives, Schlesinger Library, Radcliffe Institute, Harvard University, Cambridge, MA.

14. HBT college recommendation by Ruth Coit, headmistress, Cambridge School (June 12, 1917), Radcliffe College Student Files.

15. *Morse Journals* 23 (August 7–8, 1917).

16. Frank W. Taussig to HBT, autumn 1917, family correspondence, box 29, Taussig papers.

17. *Morse Journals* 26 (September 2, 1918): 6–7.

18. *Morse Journals* 29 (June 30, 1919): 38–40.

19. The bacille Calmette-Guerin (BCG) vaccine did not prevent the disease or stop those with antibodies from becoming infected, though it is still widely used in poor countries. Doctors began using antibiotics to treat TB shortly after these new drugs became widely available in 1949.

20. Susan J. Baserga, "The Early Years of Coeducation at the Yale University School of Medicine," *Yale Journal of Biology and Medicine* (1980):182.

21. Mary Roth Walsh, *Doctors Wanted: No Women Need Apply* (New Haven, CT: Yale University Press, 1977), 185.

22. See Walsh, *Doctors Wanted*, 181, and Mary Ann C. Elston, "Women Doctors in the British Health Services: A Sociological Study of Their Careers and Opportunities" (PhD diss., University of Leeds, 1986). Walsh reported 258 British women doctors using British census data, but these did not include women doctors serving in India and other British colonies with the support of Queen Elizabeth. A 1909 count including missionaries found 848 women doctors. Elston concluded that "there is no simple answer" for the number of British women doctors from the late 1800s to the 1980s.

23. Francis D. Moore and Cedric Priebe, "Board-Certified Physicians in the United States, 1971–1986," *New England Journal of Medicine* 324, no. 8 (February 21, 1991): 542. See also Walsh, *Doctors Wanted*, 186.

24. Paul Starr, *The Social Transformation of American Medicine* (New York: Basic Books, 1982), 155.

25. Walsh, *Doctors Wanted*, 169.

26. Kathleen Waters Sander, *Mary Elizabeth Garrett: Society and Philanthropy in the Gilded Age* (Baltimore, MD: Johns Hopkins University Press, 2008), 176–182.

27. Walsh, *Doctors Wanted*, 193.

28. Walsh, *Doctors Wanted*, 241.

29. Walsh, *Doctors Wanted*, 193.

30. Regina Morantz-Sanchez, *Sympathy and Science: Women Physicians in American Medicine* (Chapel Hill: University of North Carolina Press, 1985), 253.

31. Starr, *The Social Transformation of American Medicine*, 93–112.

32. The senior Taussig, who treated Gen. Ulysses S. Grant during the Civil War, distinguished himself thereafter as few Americans would, collecting taxes for President Lincoln, building the St. Louis bridge, sparring with Andrew Carnegie, and opening a school for the blind. William Taussig did so well that he financed the building of a substantial house at 2 Scott Street for son Frank and daughter-in-law Edith as a wedding present.

33. A. McGehee Harvey, "Helen Brooke Taussig," in *Adventures in Medical Research*, ed. A. McGehee Harvey (Baltimore, MD: Johns Hopkins University Press, 1976), 138.

CHAPTER 3

1. Helen B. Taussig interview by Charles A. Janeway, 4, Family Planning Oral History Project Interviews, August 16, 1975, OH-1; T-25; M-138; A1–3, Schlesinger Library, Radcliffe Institute, Harvard University, Cambridge, MA. Hereafter referred to as the Janeway interview. Janeway, who studied in Helen's clinic, was physician-in-chief at Boston Children's Hospital from 1946 to 1974.

2. There were several dozen homeopathic colleges in the United States. Boston University had absorbed the old New England Female Medical College, which, until its merger in 1873, had prepared women to be midwives. Following the merger, more than 30 percent of the students were women.

3. Harvey, "A Conversation with Helen Taussig," 31.

4. Helen B. Taussig and Faith L. Meserve, "Rhythmic Contractions in Isolated Strips of Mammalian Ventricles," *American Journal of Physiology* 72 (1925): 89–98.

5. Janeway interview, 5.

6. Helen Sinclair Pittman, president of Smith College, October 8, 1975, remarks in advance of a viewing of her taped interview with Helen B. Taussig for "Leaders in American Medicine," Oral History Series, Boston, Helen Sinclair Pittman Papers, correspondence, box 975, folder 9, Smith College Archives, Northampton, MA.

7. Katherine L. (Kitty) Esterly, MD, Delaware neonatologist, interview by the author, February 8, 2011. Esterly (1925–2014) was Handy's deputy and in later years her driver on trips to Baltimore and Chadds Ford, Pennsylvania.

8. Helen stood between 5'10" and 5'11" based on interviews with colleagues who compared their heights to hers, and Helen's description of a 5'9½" patient as "not quite as tall as I am."

9. Florence Sabin to HBT, April 2, 1925, Florence Rena Sabin Papers, American Philosophical Society, Philadelphia, series I, correspondence, Helen B. Taussig (1925–1936), box 22. Hereafter referred to as the Sabin papers. "I have a place for you to work and will explain to you whenever you can come in," Sabin wrote.

10. Harry O. Veach to HBT, October 17, 1974, correspondence, box 9, Taussig papers.

11. Robert A. Harrell, MD, "History of the Club," The Society of Pithotomists, accessed March 20, 2023, http://www.pithotomy.com/history.html.

12. Morantz-Sanchez, *Sympathy and Science*, 137. She calls women doctors an exceptional group: those who married were too invested in their careers to give them up and those who remained single were willing to lead unconventional personal lives.

13. One-third of women doctors who began Radcliffe with Helen in 1919 married. P. A. Williams, "Women in Medicine: Some Themes and Variations," *Journal of Medical Education* 46, no. 7 (July 1971): 584–591.

14. *Washington Post*, April 16, 1972, general correspondence, box 7, Taussig papers. "A person can't do everything she wants to in life and I've had other experiences that I wouldn't have had if I had married," Helen said. "Besides, I think I've been more of a success with what I've done than I would have been with marriage."

15. HBT to Anne Kennedy, February 18, 1926, American Birth Control League Records, series I, correspondence, 1919–28, folder 321, Houghton Library, Harvard College Library, Cambridge, MA. Kennedy, executive secretary of the league, accompanied Sanger to Helen's boardinghouse.

16. Frank Taussig to Gerhart von Schulze-Gaevernitz (June 1, 1925), Taussig and Cotuit home files, box 12, folder 1914–28. Papers of Frank W. Taussig, Harvard University Archives, Harvard University, Cambridge, MA. Hereafter referred to as the Frank W. Taussig papers. Courtesy of the Harvard University Archives.

CHAPTER 4

1. Edwards A. Park et al, *The Harriet Lane Home: A Model and a Gem* (Baltimore, MD: Department of Pediatrics, School of Medicine, Johns Hopkins University, 2006), 108.

2. Daniel M. Fox, *Power and Illness: The Failure and Future of American Health Policy* (Berkeley: University of California Press, 1993), 39–50.

3. William N. Evans, "Helen Brooke Taussig and Edwards Albert Park: The Early Years (1927–1930)," *Cardiology in the Young* 20, no. 4 (August 2010): 392. Park wrote to Helen in New York on January 23, 1929.

4. HBT to Edwards A. Park, January 28, 1929, correspondence, box 13, Taussig, H.B, folder 1929–30, Edwards A. Park Collection, Chesney Archives—Johns Hopkins Medicine, Nursing and Public Health, Baltimore, MD. Hereafter referred to as the Park papers.

5. Frank W. Taussig to Cornelia and Ellen Fisher, sisters of his late wife, Laura, January 12, 1930, Taussig and Cotuit home files, box 12, folder 1930–31, Frank W. Taussig papers.

6. HBT to Edwards A. Park, April 22, 1930, correspondence, box 13, Taussig, H.B, folder 1929–30, Park papers.

7. Evans, "Helen Brooke Taussig and Edwards Albert Park: The Early Years (1927–1930)," 393.

8. Edwards A. Park to HBT, April 12, 1930, correspondence, box 13, Taussig, H.B, folder 1929–30, Park papers.

9. HBT to Edwards A. Park, April 29, 1930, correspondence, box 13, Taussig, H.B, folder 1929–30, Park papers.

10. HBT, "Little Choice and Stimulating Environment," *Journal of the American Medical Women's Association* 36, no. 2 (February 1981): 43.

11. HBT, "Little Choice and Stimulating Environment," 43–44.

12. Rose Strauss, Helen's lipreading teacher, interview by the author, February 11, 2011. Strauss provided records of Helen's hearing loss.

CHAPTER 5

1. Peter C. English, *Rheumatic Fever in America and Britain* (New Brunswick, NJ: Rutgers University Press, 1999), 93.

2. English, *Rheumatic Fever in America and Britain*, 24.

3. Funeral program for Amy Clark, Sharp Street Memorial United Methodist Church, September 25, 1967, biographical materials and memorabilia, box 174, Taussig papers.

4. Lynn Gilbert and Gaylen Moore, *Particular Passions: Talks with Women Who Have Shaped Our Times* (New York: Clarkson N. Potter, 1981), 57.

5. Helen B. Taussig, "How to Adjust to Deafness, Hints Based on Personal Experience," *Medical Times* 109 (February 1981): 43s.

6. HBT note in patient file, November 23, 1932, rheumatic fever patient records, Taussig papers.

7. English, *Rheumatic Fever in America and Britain*, 137. Helen B. Taussig, "Acute Rheumatic Fever: Significance and Treatment of Various Manifestations," *Journal of Pediatrics* 14 (1939): 581.

8. HBT notes on Dottie Worthington (pseudonym), November 10, 1933, patient records, rheumatic fever, box 100, Taussig papers.

9. HBT note, August 10, 1933, patient records, rheumatic fever, box 103, Taussig papers.

10. HBT note, December 9, 1932, patient records, box 102, Taussig papers.

11. Raymond Gebhardt (pseudonym) to HBT, April 7, 1972, patient correspondence, box 30, Taussig papers.

12. HBT, "Acute Rheumatic Fever," 583.

13. HBT to Edwards A. Park, October 20, 1931, series I, correspondence, Helen B. Taussig, box 13, folder 1931, Park papers.

14. HBT notes on Dottie Worthington (pseudonym).

15. Frank W. Taussig, November 9, 1933, correspondence, box 12, Taussig folder, 1932–37, Frank W. Taussig papers.

16. HBT to Edwards A. Park, June 19, 1933, correspondence box 13, Helen B. Taussig, folder 1933–36, Park papers.

17. HBT notes on Dottie Worthington (pseudonym).

18. HBT, "Acceptance of the Howland Award," *Pediatric Research* 5 (October 1971): 577.

CHAPTER 6

1. Robert Collis, "The Chief," *Journal of Pediatrics* 41, no. 6 (December 1952): 664–672.

2. Janeway interview, 16.

3. HBT notes, undated, patient records, box 101, Taussig papers.

4. HBT to Edwards Park, October 1, 1934, correspondence, box 13, Taussig folder 1933–36, Park papers.

5. Frank W. Taussig to Edwards Park, November 1, 1934, Taussig and Cotuit home files, box 12, folder 1934–35, Frank W. Taussig papers.

6. Helen B. Taussig, "The Management of Children with Rheumatic Heart Disease (Compensated and Decompensated)," *Medical Clinics of North America* 18 (May 1935): 1559–1578.

7. Janeway interview, 17.

8. D. R. Barry and D. H. Isaac, "A Case of Cor Triloculare Biatriatum with Survival to Adult Life," *British Medical Journal* 2 (October 24, 1953): 921.

9. Janeway interview, 17.

CHAPTER 7

1. Helen B. Taussig, "Two Cases of Congenital Malformation of the Heart due to Defective Development of the Right Ventricle," *Bulletin of the International Association of Medical Museums* 16 (1936): 67.

2. Helen B. Taussig, "The Clinical and Pathological Findings in Congenital Malformation of the Heart due to Defective Development of the Right Ventricle Associated with Tricuspid Atresia or Hypoplasia," *Bulletin of the Johns Hopkins Hospital*, no. 59 (1936): 435–445.

3. William N. Evans, "The Relationship between Maude Abbott and Helen Taussig: Connecting the Historical Dots," *Cardiology in the Young*, no.18 (December 2008): 557–564.

4. HBT to Florence Sabin, April 15, 1936, Sabin papers.

5. Florence Sabin to HBT, April 16, 1936, Sabin papers.

6. HBT to Florence Sabin, April 18, 1936, Sabin papers.

7. Meg Fairfax Fielding, "Marcia, Marcia, Marcia," *MedChi Archives Blog*, Maryland State Medical Society, December 12, 2013, https://medchiarchives.blogspot.com/search?q=marcia+.

8. Patricia J. F. Rosof, "The Quiet Feminism of Dr. Florence Sabin: Helping Women Achieve in Science and Medicine," *Gender Forum* 24 (2009): 37.

9. Frank Taussig's wealth came partly from the company Helen's grandfather had created to build the St. Louis Bridge. In dividends alone in 1934 from the Union Pacific Railroad and Savannah Sugar Refining, he earned $4,763, the equivalent of $96,573 in 2021 dollars.

10. Edwards A. Park to HBT, June 5, 1936, box 13, folder 1933–36, Park papers.

11. Edwards A. Park to HBT, October 9, 1936, box 13, folder 1933–36, Park papers.

12. Kenneth Blackfan to Edwards A. Park, February 5, 1930, and Edwards A. Park to Kenneth Blackfan, May 28, 1937, correspondence, box 2, Park papers.

13. HBT to Edwards A. Park (summer 1937), box 13, folder 1933–36, Park papers.

14. Frank W. Taussig to Franklin Delano Roosevelt, December 29, 1936, correspondence, box 11, folder "T"-Taussig, Frank W. Taussig papers.

15. HBT to Edwards A. Park, July 28, 1937, box 13, folder 1937–38, Park papers.

16. HBT to Edwards A. Park, summer 1937, box 13, folder 1937–38, Park papers.

17. Helen B. Taussig and M. S. Hecht, "Studies Concerning Hypertension in Child-hood," *Bulletin of the Johns Hopkins Hospital*, no. 62 (1938): 482–490, 491–521.

18. HBT to Edwards A. Park July 19, 1937, box 13, folder 1937–38, Park papers.

19. Helen B. Taussig, A. McGee Harvey, and Richard H. Follis Jr., "The Clinical and Pathological Findings in Interauricular Septal Defects: A Report of Four Cases," *Bulletin of the Johns Hopkins Hospital*, no. 63 (1938): 61–89.

20. Denton Cooley interview by John Schatz, MD, May 16, 2001, transcript provided to the author.

21. Helen Brooke Taussig, *Congenital Malformations of the Heart*, 2nd ed., vol. 2 (Cambridge, MA: Harvard University Press, 1960), 23.

22. William B. Bean, MD, to HBT, October 23, 1970, correspondence, box 26, Taussig papers.

CHAPTER 8

1. Edwards A. Park to HBT, April 26, 1938, box 13, folder 1937–39, Park papers.

2. James H. Semans and Helen B. Taussig, "Congenital Aneurysmal Dilation of the Left Auricle," *Bulletin of the Johns Hopkins Hospital* 63 (1938): 404–413.

3. HBT to Edwards A. Park, July 30, 1938, box 13, folder 1937–39, Park papers.

4. Helen won permission from Park to hire a doctor working temporarily with Thomas Lewis in London. Marcel Goldenberg had been a star at the University of Vienna. Back and forth they went over visas, timetables, until, finally, it appears a US quota on German immigration may have barred Goldenberg from work at Hopkins.

5. HBT, "History of the Blalock-Taussig Operation and Some of the Long-Term Results on Patients with a Tetralogy of Fallot," First James Bordley III Lecture, February 27, 1970, Cooperstown, New York, 17.

6. Janeway interview, 19.

7. HBT, "Acute Rheumatic Fever."

8. Roger A. Crane to Roderick Heffron, October 22, 1958, Commonwealth Fund Collection, Rockefeller Archive Center, Sleepy Hollow, NY.

9. HBT to Edwards A. Park, June 29, 1939, box 13, folder 1937–39, Park papers.

10. Marjorie Prichard to HBT, August 26, 1939, correspondence with individuals, box 7, folder 20, Taussig papers.

11. Francis D. Moore and Judah Folkman, "Robert Edward Gross 1905–1988," biographical memoir (Washington, DC: National Academy of Sciences, 1985): 133,

http://www.nasonline.org/publications/biographical-memoirs/memoir-pdfs/gross
-robert.pdf.

12. Janeway interview, 21.

13. W. Hardy Hendren and M. Judah Folkman, "Robert Edward Gross," Faculty of
Medicine, Harvard University, accessed March 20, 2023, https://fa.hms.harvard.edu
/files/memorialminute_gross_robert_e.pdf.

14. Janeway interview, 21.

15. Prichard to HBT, January 27, 1941, Taussig papers.

16. Janeway interview, 21.

CHAPTER 9

1. Prichard to HBT, January 27, 1941, Taussig papers.

2. Patient's father to HBT, February 1971, patient correspondence, Taussig papers.

3. Prichard to HBT, December 19, 1939, Taussig papers.

4. Alfred E. Barclay, Kenneth J. Franklin, and Marjorie M. L. Prichard, *The Foetal
Circulation and Cardiovascular System and the Changes That They Undergo at Birth*
(Oxford: Blackwell, 1944).

5. Edwards A. Park to AB, February 26, 1941, box 15, personal and professional correspondence, 1931–1985, Alfred Blalock Papers, 1899–1985, Duke University Medical Center Archives, Chapel Hill, NC. Hereafter referred to as the Blalock papers.

6. Alfred Blalock, "Cardiovascular Surgery, Past and Present," *Journal of Thoracic and
Cardiovascular Surgery* 51, no. 23 (February 1966): 156.

7. Vivien Thomas, *Partners of the Heart* (Philadelphia: University of Pennsylvania
Press, 1985), 16.

8. Moore and Folkman, "Robert Edward Gross 1905–1988," 134.

9. Janeway interview, 22.

CHAPTER 10

1. Shirley Rosenthal entry in Barbara Rosenthal's baby book, Paul Rosenthal family
papers. Hereafter called the Rosenthal papers.

2. Emma Lodge to UCLA memorial fund for Barbara Rosenthal, Rosenthal papers.

3. Harvey, "A Conversation with Helen Taussig," 35.

4. Harvey, "A Conversation with Helen Taussig," 35.

5. Janeway interview, 23.

6. Edwards A. Park, foreword to *Congenital Malformations of the Heart*, by Helen B. Taussig (New York: Commonwealth Fund, 1947), vii.

7. Janeway interview, 24. "If you could put the carotid artery into the descending aorta, couldn't you put the subclavian artery into the pulmonary artery?" she recalled asking Blalock. See also Bordley Lecture, 18.

8. Edwards A. Park to Mark M. Ravitch, February 1965, March 4, 1965, and September 30, 1965, Mark M. Ravitch Papers, Series 5: Writing by Ravitch, 1930–1988, books, 8, Alfred Blalock-1899-1964, box 36, folders 4 and 6, National Library of Medicine, Bethesda, Maryland. Hereafter referred to as the Ravitch papers.

9. AB to HBT, September 22, 1945, books, 8, box 36, folder 6, Ravitch papers. See also Mark M. Ravitch, ed., *The Papers of Alfred Blalock* (Baltimore, MD: Johns Hopkins Press, 1966), xlv. Ravitch, quoting from the September 22, 1945, letter, said Blalock put it "pithily," when he told Helen, "I must say that if I make the statement to you that you could improve the condition of patients with aortic stenosis if you could find a means to allow more blood to reach the body, that I would be far from solving the practical problem."

10. Knut H. Leitz and Gerhard Ziemer, "The History of Cardiac Surgery," in *Cardiac Surgery, Operations on the Heart and Great Vessels in Adults and Children*, ed. Gerhard Ziemer and Axel Haverich (Berlin: Springer-Verlag, 2017), 6. Halsted's visit is mentioned in Ira M. Rutkow and Karl Hempel, "An Experiment in Surgical Education—the First International Exchange of Residents: The Letters of Halsted, Kuttner, Heuer, and Landois, January 1988," *Archives of Surgery* 123, no. 1 (1988): 115–121.

11. Thomas, *Partners of the Heart*, 80.

12. Thomas, *Partners of the Heart*, 81.

13. Helen B. Taussig, "The Development of the Blalock-Taussig Operation and Its Results 20 Years Later," *Proceedings of the American Philosophical Society* 120, no. 1 (January 1976): 15. HBT gave a similar account in the Pittman oral interview ("I went over and made various suggestions"). Janeway interview, 24, and Bordley Lecture, 18. Vivien Thomas was recognized toward the end of Blalock's career for the extraordinary part he played in the history of surgery.

14. Vivien Thomas to Mark M. Ravitch, July 7, 1966, books, 8, box 36, folder 7, Ravitch papers.

15. Vivien Thomas interview by Peter D. Olch, April 20, 1967, General History of Medicine, Oral Histories, History of Medicine Division, National Library of Medicine, Bethesda, MD, 27. See also Thomas, *Partners of the Heart*, 80.

16. HBT, draft of notes for Paul D. White, August 2, 1968, correspondence with individuals, box 9, folder 39, and HBT to I. Ridgeway Trimble, August 29, 1972, box 9, folder 18, Taussig papers. An account by an unnamed intern in Jurgen Thorwald,

The Patients (New York: Harcourt Brace Jovanovich, 1971), 13, set the time line for experimentation as December 1942.

17. Janeway interview, 27. See also Harvey, "A Conversation with Helen Taussig," 35.

18. William P. Longmire Jr., *Alfred Blalock, His Life and Times* (privately printed, 1991), 101.

19. Longmire to Mark M. Ravitch, August 5, 1965, books, 8, box 36, folder 12, Ravitch papers.

20. HBT note, November 10, 1944, patient records, Taussig papers.

21. Janeway interview, 25. HBT said Blalock told her, "We haven't got very conclusive experiments but if you are convinced it is going to help, I am convinced I know how to do the operation."

22. Janeway interview, 25.

23. Thorwald, *The Patients*, 16.

24. Thomas, *Partners of the Heart*, 91.

25. Thorwald, *The Patients*, 18.

26. Thorwald, *The Patients*, 19.

27. HBT, address on the anniversary of the 25th anniversary of the Martinsburg, WV, children's heart clinic (1979): 5, box 15, folder Berkeley County, Martinsburg, WV, Taussig papers.

28. Prichard to HBT, February 11, 1945.

29. Janeway interview, 28.

CHAPTER 11

1. Harvey, "A Conversation with Helen Taussig," 35.

2. Harvey, "A Conversation with Helen Taussig," 36.

3. Janeway interview, 30.

4. HBT, address on the anniversary, 6.

5. Harvey, "A Conversation with Helen Taussig," 36.

6. HBT, address on the anniversary, 6.

7. HBT, address on the anniversary, 6.

8. Shirley Rosenthal crossed out the original March 5 departure date for Barbara in her scrapbook and replaced it with March 13, one day after the historic surgery was unveiled (Rosenthal papers).

9. Mary Allen Engle, "The Early Years," in "Historical Milestones: Helen Brooke Taussig (1898–1986)," by Dan G. McNamara et al., *Journal of the American College of Cardiology* 10, no. 3 (September 1987): 663.

10. Warfield Firor, interview by Peter D. Olch, March 21, 1967, General History of Medicine, Oral Histories, History of Medicine Division, National Library of Medicine, 13, https://oculus.nlm.nih.gov/cgi/t/text/text-idx?c=oralhist;cc=oralhist;rgn=main;view=text;idno=2935095r.

11. Firor interview by Olch, 14.

12. Firor interview by Olch, 14.

13. Austin Lamont to Mark M. Ravitch, November 1, 1965, books, 8, box 36, folder 12, Ravitch papers.

14. Sanford E. Leeds interview by Peter D. Olch, January 29, 1973, General History of Medicine, Oral Histories, History of Medicine Division, National Library of Medicine, Bethesda, MD, 10. See also Thomas, *Partners of the Heart*, 35–38, and Sanford E. Levy and Alfred Blalock, "Experimental Observations on the Effects of Connecting by Suture the Left Main Pulmonary Artery to the Systemic Circulation," *Journal of Thoracic Surgery* 8 (June 1939): 535. Leeds, who changed his name from Levi, was a researcher at Vanderbilt in 1936–1938 before his residency.

15. Edwards A. Park to Mark M. Ravitch, March 4, 1965, books, 8, box 36, folder 4, Ravitch papers.

16. Longmire, *Alfred Blalock*, 112. See also Mark M. Ravitch to Tinsley Harrison, May 9, 1966, books, 7, box 35, folder 28, Ravitch papers. Ravitch expressed surprise that Blalock delegated experimental work to others in Nashville.

17. Alfred Blalock and Helen B. Taussig, "The Surgical Treatment of Malformations of the Heart in Which There Is Pulmonary Stenosis or Pulmonary Atresia," *Journal of the American Medical Association* 123, no. 3 (May 19, 1945): 189–202.

18. "City Council Rejects Ban on Vivisection," *Baltimore Sun*, April 10, 1945.

CHAPTER 12

1. HBT to AB, September 14, 1945, books, 8, box 36, folder 6, Ravitch papers.

2. HBT to AB, July 30, 1945, Alfred Blalock Collection 1940–1964, correspondence, box 72, folder 10, Helen B. Taussig, 1945–1964, Chesney Archives—Johns Hopkins Medicine, Nursing and Public Health, Baltimore, MD. Hereafter referred to as the Blalock collection.

3. HBT to Lester Evans, July 30, 1945, box 19, Blalock papers. Three children had not improved.

4. AB to HBT, August 23, 1945, box 72, Blalock collection.

5. HBT to AB, August 14, 1945, box 72, Blalock collection.

6. HBT to AB, July 30, 1945, box 72, Blalock collection.

7. HBT to AB, August 17, 1945, box 72, Blalock collection.

8. AB to HBT, September 22, 1945, books, 8, box 36, folder 6, Ravitch papers.

9. HBT to AB, September 14, 1945, books, 8, box 36, folder 6, Ravitch papers.

10. HBT to AB (September 1945), books, 8, box 36, folder 6, Ravitch papers.

11. Edwards A. Park to AB, November 6, 1945, personal and professional correspondence, box 15, Blalock papers.

12. "Tetralogy of Fallot, Discussion of Two Papers," *Journal of Thoracic Surgery* 16 (1947): 241. The paper was read at the meeting of the American Association for Thoracic Surgery, Detroit, May 29–31, 1946.

13. AB to HBT, June 4, 1946, box 72, Blalock collection.

14. HBT to AB November 21, 1946, box 72, Blalock collection.

15. Helen B. Taussig and Alfred Blalock, "Observations on the Volume of the Pulmonary Circulation and Its Importance in the Production of Cyanosis and Polycythemia," *American Heart Journal* 33 (April 1947): 413–419.

16. Alfred Blalock, "The Surgical Treatment of Congenital Pulmonic Stenosis," read to surgeons at a meeting in April 1946, Hot Springs, VA. Alfred Blalock, "The Surgical Treatment of Congenital Pulmonic Stenosis," *Annals of Surgery* (November 1946): 789–885.

17. HBT to AB, December 11, 1946, box 72, Blalock collection.

18. Austin Lamont to Mark M. Ravitch, November 1, 1965, box 36, folder 12, Ravitch papers.

19. HBT to AB, March 12, 1947, box 72, Blalock collection.

20. HBT to AB, March 13, 1947, box 72, Blalock collection.

21. AB to HBT, March 21, 1947, box 72, Blalock collection.

22. HBT to AB, March 24, 1947, box 72, Blalock collection.

23. HBT to AB, April 1, 1947, box 72, Blalock collection. She told him to take it up with her boss if he had any questions.

24. AB to HBT, April 4, 1947, box 72, Blalock collection.

25. Helen B. Taussig, "Analysis of Malformations of the Heart Amenable to a Blalock-Taussig Operation," *American Heart Journal* 36, no. 3 (September 1948): 321.

26. Aline Pithon to Alfred Blalock, January 14, 1947, box 15, personal and professional correspondence, 1931–1985, Paris Trip, Blalock papers.

27. Ravitch, *Papers of Alfred Blalock*, li.

28. "Among the Arrivals on the De Grasse," *New York Times*, October 17, 1947.

CHAPTER 13

1. "Baby Facing Death Flies from West Coast to Johns Hopkins for a Heart Operation," *New York Times*, December 4, 1945.

2. "Surgeons Cure Blue Babies," *Baltimore Sun*, December 8, 1945.

3. Thorwald, *The Patients*, 26. Thorwald's account has Mason describing Blalock's parents, but his father, George Z. Blalock, a merchant and cotton farmer, died in 1932.

4. Price Day, "Taussig-Blalock and Blue Babies," *Baltimore Sun Magazine*, February 3, 1946.

5. Lester Grant, "How Two Doctors Give New Lives to Blue Babies—Blalock-Taussig Operation, First Tried on Dogs, Reroutes Flow of Blood," *New York Herald Tribune*, February 15, 1946.

6. Robert D. Potter, "Saving Our Doomed Blue Babies," *Cleveland Plain Dealer*, February 17, 1946.

7. HBT letter to Carolyn Lytle, Montecito School for Girls, Santa Barbara, CA, March 3, 1950, correspondence with individuals, box 5, folder 22. Taussig papers.

8. "Dr. Helen Taussig Saves 'Blue Babies,'" *Boston Sunday Post*, October 17, 1948.

9. Thomas, *Partners of the Heart*, 98.

10. Karen Kozlowski and Meg Cohen Ragas, *Read My Lips: A Cultural History of Lipstick* (San Francisco: Chronicle Books, 1998).

11. "Surgery Can Save 'Blue Children,'" *Collier's*, April 6, 1946.

12. Longmire, *Alfred Blalock*, 111.

13. AB to HBT, April 8, 1946, box 72, Blalock collection.

14. HBT to AB, April 27, 1946, box 72, Blalock collection.

15. Geneviève Maurin, *Blue Baby, cet enfant qui ne devait vivre* (Lisieux, France: Éditions Moriere, 1949), Chesney Archives—Johns Hopkins Medicine, Nursing and Public Health, Baltimore, MD. Translated for the author by Annette Cripps.

CHAPTER 14

1. Janeway interview, 32.

2. HBT, Bordley Lecture, 27.

3. HBT to Louis Flexner, June 18, 1945, correspondence, box 59, folder 12, pulmonary stenosis, Blalock collection.

4. Besides Cournand's lab at Bellevue Hospital, Lewis Dexter in Boston used the blood sampling method in his lab. Cournand and Richards shared the 1956 Nobel

Prize with Werner Forssmann of Germany, who in 1929 first inserted a catheter into his own vein to test its usefulness.

5. Richard J. Bing, "The Johns Hopkins: The Blalock-Taussig Era," *Perspectives in Biology and Medicine* 32, no. 1 (Autumn 1988): 85–90.

6. Igor E. Konstantinov, "Taussig-Bing Anomaly: From Original Description to the Current Era," *Texas Heart Institute Journal* 36, no. 6 (2009): 584.

7. Konstantinov, "Taussig-Bing Anomaly," 583.

8. HBT to AB, March 24, 1947, box 72, Blalock collection.

9. Thomas, *Partners of the Heart*, 105, and Vivien Thomas interview by Peter D. Olch, 18, Oral Histories, National Library of Medicine.

10. HBT, "Analysis of Malformations of the Heart Amenable to a Blalock-Taussig Operation," 325.

11. After catheterization became safer, physicians injected contrasting medium (dye) into the arteries and watched the blood flow on camera to diagnose transposed vessels.

12. HBT, "On the Evolution of Our Knowledge of Congenital Malformations of the Heart," T. Duckett Jones Memorial Lecture, *Circulation* 31 (May 1965): 774.

13. Park, foreword to *Congenital Malformations of the Heart*, viii.

14. Helen B. Taussig and Richard J. Bing, "Complete Transposition of the Aorta and a Levoposition of the Pulmonary Artery," *American Heart Journal* 37, no. 4 (1949): 551.

15. Charlotte Ferencz, interview by the author, December 2010.

16. Bing's early contributions to heart disease may have been overlooked because they were not published in journals read by pediatric cardiologists. See Dan G. McNamara, "Contributions of Richard Bing to the Field of Congenital Heart Disease," *Journal of Applied Cardiology* 4 (1989): 354. Hopkins in 2000 awarded Bing an honorary degree, citing his path-breaking description of heart defects. He authored four hundred scientific articles and three hundred music works.

17. Sissman, interview by the author, February 25, 2011.

18. Dotter, longtime chair of the Department of Radiology at the University of Oregon Medical School and founder of interventional radiology, invented new material for the catheter so it could be used as a safer intervention than surgery.

19. Heinrich Taegtmeyer, interview by the author, November 26, 2013.

20. Mary Ellen Avery interview by Lawrence M. Gardner, MD, April 4, 1998, Pediatric History Center, Oral History Project, American Academy of Pediatrics, Elk Grove Village, IL, 19, https://www.aap.org/en/about-the-aap/gartner-pediatric-history-center/oral-histories/?facets=%5B%5D&k=mary%20ellen%20avery&page=1.

21. HBT Lillian Welsh Lecture, 7, April 3, 1954, biographical materials, box 44, Taussig papers.

22. HBT, Lillian Welsh Lecture, 20. Barringer was the sister-in-law of Rose Morse, the daughter of Cotuit artist Lucy Gibbons Morse and the diarist James Morse. Helen visited her in New Canaan, CT, on road trips in the 1950s.

23. Helen V. Crouse, "Perspectives in Science," *Goucher Alumnae Quarterly* 32, no. 3 (Spring 1954): 3.

24. Bing's scolding sent Charlotte in tears to the Garrett Room, a meeting space for women doctors financed by M. Carey Thomas after Garrett's death. Bing later made amends, sending Charlotte a massive bouquet of flowers.

CHAPTER 15

1. HBT, *Congenital Malformations of the Heart*, ix.

2. F. N. Low, "New Biological Books," *Quarterly Review of Biology* 23, no. 4 (December 1948): 365, http://www.journals.uchicago.edu/doi/abs/10.1086/396684.

3. William J. Potts, a Chicago surgeon, believed it safer to suture the aorta directly to the pulmonary vessel. His technique did not improve long-term outcomes, however, and was used only in difficult cases.

4. HBT to Charlotte Ferencz, July 18, 1952, correspondence with Helen Taussig, box 2, Charlotte Ferencz Collection, Chesney Archives—Johns Hopkins Medicine, Nursing and Public Health, Baltimore, MD. Hereafter referred to as the Ferencz papers.

5. Richard S. Ross, interview by the author, May 11, 2013. Ross met Helen as an intern in 1947, worked at her side as chief medical resident in 1953, developed nonsurgical treatments for heart patients, and served as Hopkins dean of medicine from 1975 to1990.

6. Taussig drawing, "My Land in Baltimore County 1948–77," biographical materials, box 174, Taussig papers.

CHAPTER 16

1. Ruth Whittemore, "Reflections of a Harriet Lane Cardiac Fellow on the First Years After the Blalock-Taussig Report," in "Historical Milestones: Helen Brooke Taussig (1898–1986)," by Dan G. McNamara et al., *Journal of the American College of Cardiology* 10, no. 3 (September 1987): 664.

2. Anthony "Tony" Perlman, interview by the author, February 11, 2011.

3. Sissman, interview by the author.

4. Records from Helen's lipreading tutor, Rose Strauss, provided to the author.

5. Richard S. Ross, presentation of the George M. Kober medal (posthumously) to Helen B. Taussig, Transactions of the Association of American Physicians, vol. C, 1987, cxix, Taussig papers.

6. Robert I. Levy, interview by the author, 2011. In a practice of fifty years, Levy, a children's kidney specialist and head of nephrology at Sinai Hospital, Baltimore, always examined children first with his hands before the stethoscope.

7. Lawrence K. Altman, "A Transplant Surgeon Who Fears Surgery," *New York Times*, July 7, 1992.

8. Francis Milligan, interview by the author, July 2, 2012.

9. Henry A. Kane, interview by the author, February 14, 2011.

10. M. A. Engle, "Ventricular Septal Defect in Infancy," *Pediatrics*, no. 14 (1954): 16–27.

11. Patsy Perlman, interview by the author, February 11, 2011. Patsy, the daughter of the pathologist Ella Oppenheimer, met Helen as a child.

12. Marie Brown interview by John Schatz, MD, October 16, 2002, provided to the author.

13. Park et al., *The Harriet Lane Home*, 233.

14. Charlotte Ferencz, interview by the author.

15. Fellows under Taussig 1946–66, corporate correspondence, box 17, folder 17, Taussig papers.

CHAPTER 17

1. Maria Blackburn, "Under Surgery's Yoke," *Hopkins Medicine* (Fall 2009).

2. Donald F. Proctor, "Experiences with Alfred Blalock and Anesthesiology at Johns Hopkins," January 1968, 8, Warfield Firor Papers, 1965–1968, Modern Manuscripts Collection, History of Medicine Division, National Library of Medicine, Bethesda, MD; MS C 195. Hereafter referred to as the Firor papers.

3. Denton A. Cooley and O. H. Frazier, "The Past 50 Years of Cardiovascular Surgery," *Circulation* 102 (November 14, 2000): IV-88.

4. HBT, Bordley Lecture, 24.

5. Mary Allen Engle interview, in *Heart to Heart: The Twentieth Century Battle Against Cardiac Disease: An Oral History*, by Allen B. Weisse (New Brunswick, NJ: Rutgers University Press, 2002), 43. She rotated through surgery under Cooley.

6. Denton A. Cooley, *100,000 Hearts: A Surgeon's Memoir* (Austin: Dolph Briscoe Center for American History/University of Texas Press, 2012), 56.

7. AB to HBT, January 3, 1950, box 72, Blalock collection.

8. Longmire, *Alfred Blalock*, 114.

9. Longmire, *Alfred Blalock*, 115.

10. Thomas, *Partners of the Heart*, 122. He was assisted by a student, Rowena Spencer, the lone woman surgical intern under Blalock.

11. HBT to AB, July 13, 1949, Blalock collection.

12. Alfred Blalock and C. Rollins Hanlon, "The Surgical Treatment of Complete Transposition of the Aorta and the Pulmonary Artery," *Surgery, Gynecology, and Obstetrics* 90 (January 1950).

13. "Blue Baby Research," *Life*, March 14, 1949, 105.

14. Cooley, *100,000 Hearts*, 75.

15. Cooley, *100,000 Hearts*, 75.

16. The first use of a crude defibrillator to revive a patient in cardiac arrest was by Claude Beck at Case Western Reserve University in 1947, but the idea of defibrillation, or sending an electrical shock to the heart to revive and reset its normal rhythms, had been under study for years at Hopkins by a team led by Kouwenhoven.

17. Proctor, "Experiences with Alfred Blalock and Anesthesiology at Johns Hopkins," 11, and addendum, a document asserting the authority of anesthesiologists over surgeons, "Division of Anesthesiology, The Johns Hopkins Hospital and School of Medicine, Prepared in January and February 1955," 1–3, Firor papers. Among Proctor's concerns was whether anesthesiologists would be legally responsible if their decisions were overridden by surgeons.

18. HBT, "Congenital Malformations of the Heart: The Clinician's Responsibility in the Selection of Patients for Operations," *Journal of Pediatrics* 41 (1952): 853.

19. Horn, interview by the author.

20. Strauss, notes provided to the author.

CHAPTER 18

1. *Baltimore Sun*, January 2, 1946. See also "Dr. Taussig Calls for Help in War on Rheumatic Fever," *Baltimore Sun*, February 13, 1944. Helen pleaded for more beds twice in 1946.

2. Joseph Ferretti and Werner Köhler, "History of Streptococcal Research," in J. J. Ferretti, D. L. Stevens, and V. A. Fischetti, eds., *Streptococcus pyogenes: Basic Biology to Clinical Manifestations* (Oklahoma City: University of Oklahoma Health Sciences Center, 2016), https://www.ncbi.nlm.nih.gov/books/NBK333430/pdf/Bookshelf_NBK333430.pdf.

3. Milton Markowitz interview by Howard A. Pearson, July 17, 1998, oral history project, Pediatric History Center, American Academy of Pediatrics, Elk Grove Village, IL, 16.

4. Markowitz interview by Pearson, 16.

5. HBT to Charlotte Ferencz, November 26, 1953, box 2, Ferencz papers.

6. Wannamaker, later professor of pediatrics and the chief of pediatric infectious diseases at the University of Minnesota, won the 1954 Albert Lasker Award for his discovery. At the time, he estimated as many as 250,000 new cases of rheumatic fever appeared annually, and 450,000 people suffered from heart disease caused by rheumatic fever.

7. He quit his practice and, as head of the pediatrics department at Baltimore's Sinai Hospital, which provided him special labs, he continued his studies. After Charlotte, he also headed Helen's rheumatic clinic until it closed.

8. Markowitz interview by Pearson, 28.

CHAPTER 19

1. W. Bruce Fry, *Caring for the Heart: Mayo Clinic and the Rise of Specialization* (Oxford: Oxford University Press, 2015), 235.

2. Fry, *Caring for the Heart*, 197.

3. Charles Bailey interview, in *Heart to Heart: The Twentieth Century Battle Against Cardiac Disease: An Oral History*, by Allen B. Weisse (New Brunswick, NJ: Rutgers University Press, 2002), 74. Bailey told Weisse said he was inspired by Smithy and even listened to Smithy's heart and heard the terrible rumble of aortic stenosis.

4. Lorenzo Gonzalez-Lavin, "Charles P. Bailey and Dwight E. Harkin—the Dawn of the Modern Era of Mitral Valve Surgery," *Annals of Thoracic Surgery* 53 (1992): 917.

5. AB to HBT, July 31, 1948, box 72, Blalock collection.

6. AB to HBT, July 31, 1948. See also Fred A. Crawford Jr., "Horace Smithy: Pioneer Heart Surgeon," *Annals of Thoracic Surgery* 89 (2010): 2070, and Fry, *Caring for the Heart*, 236.

7. E. Cowles Andrus to Mark M. Ravitch, July 8, 1965, books, 8, box 36, folder 6, Ravitch papers. See also A. P. Naef, "The Mid-Century Revolution in Thoracic and Cardiovascular Surgery: Part 5," *Interactive Cardiovascular and Thoracic Surgery* 3, no. 3 (2004): 422, https://academic.oup.com/icvts/article/3/3/415/765692.

8. AB to HBT, October 4, 1949, box 72, Blalock collection.

9. Thomas, *Partners of the Heart*, 99. Thomas noted that Blalock operated on the first blue baby and surgeries in Nashville without practice, though he had previously stitched together two blood vessels.

10. Cooley, *100,000 Hearts*, 74.

11. Catherine A. Neill and Edward B. Clark, *The Developing Heart: A "History" of Pediatric Cardiology* (Dordrecht: Springer Science + Business Media, 1995), 122. Blalock in

1946 also wrote Helen about a case in which he wrapped an artery with cellophane, but the patient died.

12. Vincent L. Gott, "C. Walton Lillehei and Total Correction of Tetralogy of Fallot," *Annuals of Thoracic Surgery* 49, no. 2 (February 1990): 330. See also Richard C. Daly et al., "Fifty Years of Open-Heart Surgery at the Mayo Clinic," *Mayo Clinic Proceedings* 80, no. 5 (1990): 636–640.

13. AB to HBT, March 27, 1950, Blalock papers.

14. Cooley, *100,000 Hearts*, 99.

15. Fry, *Caring for the Heart*, 203.

16. Henry T. Bahnson oral interview by William S. Stoney, Annette and Irwin Eskind Biomedical Library, Vanderbilt University Medical School, August 3, 1999, https://www.library.vanderbilt.edu/specialcollections/history-of-medicine/exhibits /cardiac_surgery/bahnson.php.

17. The parent's femoral artery and vein were connected to the patient's arterial and venous system, respectively (veins carry blood to the heart; arteries carry oxygenated blood away).

18. Bahnson interview by Stoney.

19. Fry, *Caring for the Heart*, 218. Lillehei had operated on at least six tetralogy of Fallot patients at this time.

20. Bahnson interview by Stoney. Bahnson's idea was to cut into the left ventricle and insert a stent into the aortic valve between the ventricle and the aorta. When he tried this on a beating heart in 1955, the patient died. He also scoured Hutzler's department store in Baltimore for fabrics to model a synthetic flap for a malfunctioning valve and eventually pioneered valve replacements.

21. HBT to AB, December 1954, Blalock collection.

22. James P. Isaacs, "My View of Alfred Blalock from 1945–1965," 5, Firor papers.

23. Proctor, "Experiences with Alfred Blalock and Anesthesiology at Johns Hopkins," Firor papers, 12. One account in the Firor papers said the operation was cross-circulation, but Bahnson in his Vanderbilt interview said he never tried the procedure on a human.

24. Proctor, "Experiences with Alfred Blalock," 12.

25. Fry, *Caring for the Heart*, 229. Gott, in his article "C. Walton Lillehei and Total Correction of Tetralogy of Fallot," called it a DeWall Lillehei bubble oxygenator. This was also called a pump-oxygenator; it both pumped and oxygenated blood on behalf of the patient whose heart was stilled during surgery.

26. Cooley, *100,000 Hearts*, 104.

27. Daly, "Fifty Years of Open-Heart Surgery at the Mayo Clinic," 637.

28. Bahnson would build the University of Pittsburgh into a powerhouse for heart surgery.

29. Isaacs, "My View of Alfred Blalock from 1945–1965," 6.

30. Vivien Thomas interview by M. Booth and G. Gordon, October 2, 1976, Victor Almon McKusick Collection, Hopkins Medical Archives, https://soundcloud.com /hopkins-medical-archives.

CHAPTER 20

1. Charlotte Ferencz and HBT, "Microscopic Study of the Pulmonary Vascular Bed of over 100 Patients Who Died following a Systemic Pulmonary Anastomosis," (Abstract) Third World Congress of Cardiology, September 14–19, Brussels, 1958.

2. F. C. Spencer and H. T. Bahnson, "The Present Role of Hypothermia in Cardiac Surgery," *Circulation* 26 (August 1962): 298.

3. C. Julian Ormand et al., "Hypothermia in Open Heart Surgery," *AMA Archives of Surgery* 73, no. 3 (1956): 493–502.

4. Discussion of Three Papers, December 5, 1962, Boca Raton, FL; *Transactions of the Southern Surgical Association* 74 (1962): 269; in Ravitch, *The Papers of Alfred Blalock*: 1952.

5. HBT to Charles Bailey, August 22, 1959, corporate correspondence, box 19, Taussig papers.

6. HBT, "Editorial: On the Selection of Patients for Surgical Repair in Congenital Defects," *Circulation* 18, no. 3 (September 1958): 321–324.

7. Admiral Elmo Zumwalt Jr. and Lieutenant Elmo Zumwalt III, with John Pekkanen, *My Father, My Son* (New York: Macmillan, 1986), 18.

8. Zumwalt et al., *My Father, My Son*, 20.

9. HBT, "Editorial: On the Selection of Patients for Surgical Repair in Congenital Defects," 324.

10. J. Alex Haller Jr., interview by the author, November 17, 2010.

11. Blalock's willingness to redo an operation reflected his call to excellence, according to surgeon C. Rollins Hanlon, who remembered Blalock surveying a vein graft he had just installed and saying, "Let's do it over."

12. Mark M. Ravitch to A. McGehee Harvey, January 6, 1986, Vivien Thomas, box 64, folder 63, Ravitch papers.

13. Catherine Neill and Helen B. Taussig, "Indications and Contraindications for Surgery in Ventricular Septal Defect," *Journal of Pediatrics* 55 (September 1959): 374–381.

14. "Current Status of Palliative vs. Corrective Procedures in Congenital Heart Surgery," *Diseases of the Chest* 43, no. 4 (April 1963): 340. This was the report of a round table discussion at the 28th Annual Meeting, American College of Chest Physicians, Chicago, June 21–25, 1962.

15. Anne M. Murphy and Duke E. Cameron, "The Blalock-Taussig-Thomas Collaboration, a Model for Medical Progress," *JAMA* 300, no. 5 (July 16, 2008): 329.

16. Konstantinov, "Taussig-Bing Anomaly," 584.

17. HBT to Charlotte Ferencz, November 22, 1959, Ferencz papers.

18. Patient to HBT, August 20, 1974, patient correspondence, Taussig papers.

CHAPTER 21

1. Ruth Bogard Gans letter to the editor, *Baltimore Sun*, February 18, 2011.

2. HBT, "Congenital Malformations of the Heart: The Clinician's Responsibility in the Selection of Patients for Operations," 858. See also Gerri Lynn Goodman, "A Gentle Heart: The Life of Helen Taussig" (master's thesis, Yale Medicine, Digital Library 2658, 1983), 56, http://elischolar.library.yale.edu/ymtdl/2658.

3. Patient to HBT, April 7, 1972, patient correspondence, Taussig papers.

4. "The Periscope," *Newsweek*, March 18, 1957.

5. H. T. Bahnson et al., "Surgical Treatment and Follow-Up of 147 Cases of Tetralogy of Fallot Treated by Correction," *Journal of Thoracic and Cardiovascular Surgery* 44 (October 1962): 419–432.

6. HBT to parent of patient, September 29, 1966, patient correspondence, box 32, folder 3, Taussig papers.

CHAPTER 22

1. HBT to Charlotte Ferencz, March 30, 1960, box 2, Ferencz papers.

2. George Martin, *CCB: The Life and Century of Charles C. Burlingham* (New York: Hill and Wang, 2005), 519. Helen, in a letter to CCB, told him she suspected he was behind her nomination.

3. Lulu Hartarian to Charlotte Ferencz, January 24, 1962, Ferencz papers.

4. Widukind Lenz, *Lancet*, February 3, 1962, box 73, Taussig papers.

5. "To Observe . . . to Deduce . . . to Know," acceptance speech by HBT for the American Association of University Women Achievement Award, *AAUW Journal* (October 1963): 25, box 45, Taussig papers.

6. HBT to A. J. Beuren, February 6, 1962, thalidomide correspondence, box 34, Taussig papers.

7. Louis Lasagna, *The Doctors' Dilemma* (New York: Harper, 1962), 131.

8. Suzanne White Junod, "FDA and Clinical Drug Trials: A Short History," US Food and Drug Administration, n.d., 8, https://www.fda.gov/media/110437/download. See also David Healy, "The Tragedy of Lou Lasagna," *Dr. David Healy* (blog), April 9, 2013, https://davidhealy.org/the-tragedy-of-lou-lasagna/. Lasagna supported some version of informed consent and animal trials.

9. Louis Lasagna, "Thalidomide—a New Non-Barbiturate Sleep-Inducing Drug," *Journal of Chronic Diseases* 11, no. 6 (1960): 627–631 His study showed a small dose was no more effective than a placebo. In 200 mg doses, he found it worked, with a few "hangover" side effects. He called it worth trying for those who did not respond to older drugs but more experience was needed to determine how high a dose was safe in humans.

10. A. Leslie Florence, "Is Thalidomide to Blame?," *BMJ* 2, no. 5217 (December 31, 1960): 1954.

11. Frances O. Kelsey, interview by the author, July 23, 2011. At age ninety-seven, she worked a crossword, whiskey in hand, daily.

12. Rock Brynner and Trent Stephens, *Dark Remedy: The Impact of Thalidomide and Its Revival as a Vital Medicine* (Cambridge, MA: Perseus, 2001), 48.

13. Frances O. Kelsey memo, September 7, 1961, exhibit 11, Chronology and Excerpts from the Official Food and Drug Administration File on the New Drug Application "Kevadon," Hearings on Interagency Coordination in Drug Research and Coordination, Senate Committee on Government Operations, Subcommittee on Reorganization and International Organizations, US Senate, 87th Congress, August 1 & 9, 1962, Part 1 (Washington, DC: US Government Printing Office, 1963), 96.

14. Janeway interview, 60.

15. HBT to Catherine Neill, March 23, 1962, thalidomide correspondence, box 37, Taussig papers.

16. HBT to Charlotte Ferencz, April 1962, Ferencz papers.

17. HBT to Paul D. White, April 3, 1962, thalidomide correspondence, box 39, Taussig papers.

CHAPTER 23

1. Kelsey memo, April 6, 1962, exhibit 11, 97.

2. Kelsey memo, April 11, 1962, exhibit 11, 81.

3. Helen Pittman interview of Helen Brooke Taussig, Alpha Omega Alpha, Leaders in American Medicine, Oral History Series, National Library of Medicine, Atlanta, GA, March 13, 1975.

4. Robert K. Plumb, "Deformed Babies Traced to a Drug—Harmless Tablet Given Abroad Is Cited as Infants' Crippler," *New York Times*, April 12, 1962, 37.

5. "The Control of Pharmaceuticals," *New York Times*, April 13, 1962, 34.

6. HBT to Guy M. Everett, May 1, 1962, Abbott Laboratories, box 34, thalidomide correspondence, Taussig papers.

7. Now the Carnegie Institute Department of Embryology, a research institute affiliated with Johns Hopkins.

8. HBT to John Talbott, May 4, 1962, thalidomide correspondence, box 34, Taussig papers.

9. HBT, "Dangerous Tranquility," *Science* 136, no. 3517 (May 25, 1962): 683.

10. "Another Toxic Hazard," *New England Journal of Medicine* 266 (June 21, 1962): 1334.

11. HBT to Saul Jarcho, August 25, 1962, thalidomide correspondence, box 36, Taussig papers.

12. Charles U. Lowe, University of Buffalo, to HBT, May 14, 1962, thalidomide correspondence, box 36, Taussig papers.

13. Charles A. Janeway to HBT, July 23, 1962, thalidomide correspondence, box 36, Taussig papers.

14. Cynthia A. Connolly, *Children and Drug Safety: Balancing Risk and Protection in 20th Century America* (New Brunswick, NJ: Rutgers University Press, 2018), 65.

15. Morris Fishbein, "Warning Given on Interfering with Development of Products," *New York Times*, May 12, 1962.

16. Harvey, "A Conversation with Helen Taussig," 42.

17. Statement of Dr. Helen B. Taussig on HR 6245, the Drug Industry Antitrust Act, May 24, 1962, Anti-Trust Subcommittee, Committee on the Judiciary, US House of Representatives.

18. HBT to Charles U. Lowe, June 19, 1962, thalidomide correspondence, box 36, Taussig papers.

19. Morton Mintz, "What Happens When No One Is Watching?," *Neiman Reports*, Neiman Foundation at Harvard, March 20, 2009, https://niemanreports.org/articles/what-happens-when-no-one-is-watching/.

20. HBT, "A Study of the German Outbreak of Phocomelia: The Thalidomide Syndrome," *JAMA* 180 (June 30, 1962): 1106–1114.

21. Morton Mintz, "'Heroine' of FDA Keeps Bad Drugs off Market," *Washington Post*, July 15, 1962.

22. Barbara Yuncker, *New York Post* reporter, to HBT, July 29, 1962, box 77, thalidomide.

23. HBT to Ashley Montagu, July 20, 1962, thalidomide correspondence, box 37, Taussig papers.

24. HBT, "The Thalidomide Syndrome," *Scientific American* 207, no. 8 (August 1962).

25. W. S. Merrell Internal Summary, Scientific Writings, box 74, Taussig papers.

26. "Drug Babies Change People's Minds over Ending Life," *Daily Mail*, July 25, 1962.

27. HBT to Stanley Birnbaum, MD, summer 1962, scientific writings, box 79, thalidomide, Taussig papers.

28. Telegram asking HBT to wire Science Service, July 27, 1962, box 38, Taussig papers.

29. HBT to John Lear, August 6, 1962, thalidomide and drug regulation, box 62, Taussig papers. He quoted her in "The Unfinished Story of Thalidomide," *Saturday Review*, September 1, 1962.

30. Richard Oulahan Jr., "Euthanasia—Should One Kill a Child in Mercy . . . or Is Life, However Hard, Too Dear to Lose?," *Life*, August 10, 1962, 34.

31. HBT notebook of media contacts, box 79, Taussig papers.

32. HBT to John Lear, August 6, 1962.

33. Howard A Rusk, "Drug-Test Regulations; US Proposals Innocuous on Surface, But Lowering of Standards Is Feared," *New York Times*, September 16, 1962.

34. Josef Warkany to HBT, July 21, 1962, box 39, Taussig papers.

35. Neil Vargesson, "Thalidomide-Induced Teratogenesis: History and Mechanisms," *Birth Defects Research Part C* 105 (2015): 141.

36. HBT to Oren Harris, chair, House Interstate and Foreign Commerce Committee, August 30, 1962, FDA Actions 1962–1963, box 62, Taussig papers.

37. Ross, interview by the author, October 16, 2013.

38. Morris Fishbein interview by Charles O. Jackson, March 12, 1968, General History of Medicine, Oral Histories, National Institutes of Health, National Library of Medicine, 60m, https://oculus.nlm.nih.gov/cgi/t/text/text-idx?c=oralhist;idno=2935140r.

39. HBT to Marcolino G. Candau, August 18, 1962, box 62, Taussig papers.

40. HBT to Charlotte Ferencz, July 22, 1962, box 2, Ferencz papers.

41. HBT, "The Evils of Camouflage as Illustrated by Thalidomide," *NEJM* 269 (July 11, 1963): 92–94.

42. HBT to Hubert H. Humphrey, March 19, 1963, FDA actions, box 62, Taussig papers.

43. HBT drug testimony before the Monopoly Subcommittee of the US Senate Small Business Committee, November 28, 1967, on legislation to eliminate export of inferior drugs, box 63, Taussig papers.

44. HBT to Oskar Hepp, April 4, 1962, thalidomide correspondence, box 36, Taussig Papers.

45. *Modern Medicine*, January 21, 1963, box 45, Taussig papers.

46. HBT to Charlotte Ferencz, February 24, 1963, box 2, Ferencz papers.

CHAPTER 24

1. Jamie Wyeth, interview by the author, February 2011.

2. HBT to Charlotte Ferencz, August 23, 1963, box 2, Ferencz papers.

3. McNamara to Mary Jane Luke, spring 1991, Fine Art File, folder Taussig/Wyeth 1986–2000, Chesney Archives—Johns Hopkins Medicine, Nursing and Public Health, Baltimore, MD. Hereafter referred to as the Taussig/Wyeth art file.

4. Cooke, interview by the author, July 20, 2011, and Mary Allen Engle, "Taussig Portraits," April 2000, Taussig/Wyeth art file.

5. HBT to AB, April 1964, box 19, Blalock papers.

6. Nils Hansson and Thomas Schlich, "Why Did Alfred Blalock and Helen Taussig Not Receive the Nobel Prize?," *Journal of Cardiac Surgery* 30 (2015): 506–509.

7. Engle, "Taussig Portraits," Taussig/Wyeth art file.

8. HBT to Jane Sabersky, September 29, 1966, box 206, terminated grants, portrait fund, Taussig papers.

9. McNamara to Mary Jane Luke, spring 1991, Taussig/Wyeth art file.

10. Catherine Neill, "Professional Career, 1955 to 1986," in "Historical Milestones: Helen Brooke Taussig (1898–1986)," by Dan G. McNamara et al., *Journal of the American College of Cardiology* 10, no. 3 (September 1987): 665.

11. HBT to William V, Moore, executive vice president of the American Heart Association, November 29, 1976, corporate correspondence, box 14, Taussig papers. HBT sent her presidential address to Moore to quietly voice her opinion on why the group should not align itself with the AMA.

12. Peter Young, "Dr. Helen Taussig Reflects on Career as Pediatrician," *Baltimore Evening Sun*, June 19, 1963.

13. Dan G. McNamara, "Her Influence in Establishing Pediatric Cardiology," in "Historical Milestones: Helen Brooke Taussig (1898–1986)," by Dan G. McNamara et al., *Journal of the American College of Cardiology* 10, no. 3 (September 1987): 667.

14. John Dorsey, "The Evolution of Heart Surgery," *Baltimore Sun*, November 29, 1964.

15. Bahnson interview by Stoney.

16. HBT, "A Time for Waiting," *Johns Hopkins Magazine* (Spring 1969): 9–11.

17. Renee C. Fox and Judith P. Swazey, *The Courage to Fail: A Social View of Organ Transplants and Dialysis* (Chicago: University of Chicago Press, 1974), 123.

18. Commonwealth Fund, "US Healthcare from a Global Perspective, 2019: Higher Spending, Worse Outcomes?," January 30, 2020, https://www.commonwealthfund .org/publications/issue-briefs/2020/jan/us-health-care-global-perspective-2019.

19. Hearings on HR 12453, International Health Act of 1966, Committee on Interstate and Foreign Commerce, US House of Representatives (Washington, DC: US Government Printing Office, February 16, 1966), 92.

20. Janeway interview, 77.

21. HBT to Barbara Diamondstein, October 1971, box 61, abortion folder, Taussig papers.

22. "Blue Baby MD Reuniting with Early Patient," *Annapolis Capital*, February 11, 1986.

23. H. B. Taussig, "Death from Lightning and the Possibility of Living Again," *Annals of Internal Medicine* 68 (1968): 1345–1353. See also Helen B. Taussig, "Death from Lightning & the Possibility of Living Again," *American Scientist* 57, no. 3 (1969): 306–316.

24. HBT, "How to Adjust to Deafness."

25. Stan Wayman, "Concentration Camp for Dogs," *Life Magazine*, February 6, 1966.

26. American missionary Eda Scudder, one of the first women to graduate from Cornell Medical School, opened the school to train women doctors and midwives after witnessing women die in childbirth. They refused to be treated by male doctors.

27. HBT to Charlotte Ferencz, May 30, 1964, box 2, Ferencz papers.

28. Anonymous Friend to HBT, May 30, 1972, correspondence, Taussig papers.

29. HBT, "The Development of the Blalock Taussig Operation Twenty Years Later," *Proceedings of the American Philosophical Society* 120, no. 1 (1976): 18.

30. HBT, "Development of the Blalock Taussig Operation," 19.

31. HBT, "Development of the Blalock Taussig Operation," 20.

32. Stephanie Goontz, "Why Gender Equality Stalled," *New York Times*, February 17, 2013, and data from the General Social Survey, in David A. Cotter et al. "The End of the Gender Revolution? Gender Role Attitudes from 1977 to 2008," *American Journal of Sociology* 117, no. 1 (July 2011): 259–289.

33. HBT to Charlotte Ferencz, November 26, 1953, box 2, Ferencz papers.

34. Lawson Wilkins, considered the father of pediatric endocrinology, was named a full professor in 1957, two years before Helen.

35. Edwards A. Park to Mark M. Ravitch, March 4, 1965, Ravitch papers.

36. Estrellita Karsh, medical historian, interview by the author, 2013.

37. Karsh, interview by the author.

38. Yousuf Karsh, *Karsh: A 50-Year Perspective* (Boston: Little, Brown, 1983).

39. Mark Eaton, interview by the author, March 10, 2012.

40. HBT last will and testament, 1983, Chester County, PA.

41. The proceeds represented the equivalent of almost two-thirds of her estate upon her death and allowed her to make substantial bequests in honor of her parents to Harvard and Radcliffe. Hopkins benefited from the sale of her cottage, minus her advance.

42. Descendant of HBT fellow, interview by the author, June 2020. HBT was a staunch supporter of the Phoenix Society, an antebellum organization to promote education and welfare of freed Blacks.

43. HBT to Mrs. Alton Jones, January 6, 1969, correspondence with individuals, box 6, folder 12, Taussig papers. Nettie Marie Jones of Easton, Maryland, was a renowned Talbot County philanthropist.

44. David Alston, interview by the author, October 21, 2011.

45. Helen B. Taussig, "Evolutionary Origin of Cardiac Malformations," *Journal of the American College of Cardiology* 12, no. 4 (October 1988): 330–334.

46. Ross, "Presentation of the George M. Kober Medal (Posthumously) to Helen B. Taussig," *Transactions of the American Physicians* C (1987): cxxv.

47. Joyce Baldwin, "A Troubling Tribute," *American Medical News*, November 24, 1989.

48. Helen's contribution was included in the final version, in a chapter written by Harvard pediatric cardiologist Alexander Nadas.

49. Longmire, *Alfred Blalock,* 115.

50. HBT to I. Ridgeway Trimble, August 29, 1972, correspondence, box 9, Taussig papers.

51. Mary Jane Luke to Charlotte Ferencz, April 17, 1990, Ferencz papers.

52. Charlotte Ferencz to Richard S. Ross, April 2, 1991, Ferencz papers.

53. Konstantinov, "Taussig-Bing Anomaly," 580–585.

54. Elizabeth Allemang, "The Midwife-Witch on Trial: Historical Fact or Myth?," *Canadian Journal of Midwifery Research and Practice* 9, no. 1 (Spring 2010): 14. Some women doctors did endure fiery deaths at the hands of those who feared their power, but they were not the focus of such persecutions.

55. Janice E. Clemens, interview by the author, August 31, 2011.

56. Committee on Faculty Development and Gender, Johns Hopkins University School of Medicine, Final Report (November 2005): 24, https://www.hopkins medicine.org/women_science_medicine/_files/new_files/cfdgfinalreport1.pdf.

57. Edward B. Clark to Richard S. Ross, spring 2000, Taussig/Wyeth art file.

58. Edward B. Clark, interview by the author, March 10, 2011.

59. Ross to Mary Allen Engle, May 24, 2000, Taussig/Wyeth art file.

60. Patricia Meisol, "The Changing Face of a Strong Woman," *New York Times*, August 14, 2013.

61. Peter Young, "Dr. Taussig Reflects on Career as Pediatrician," *Evening Sun*, June 19, 1963.

62. "Excerpts from Dr. Taussig's Lecture," *Goucher Alumnae Quarterly* 32, no. 3 (Spring 1954): 5.

INDEX